TECHNIQUES IN
HIGH PRESSURE
NEUTRON
SCATTERING

TECHNIQUES IN
HIGH PRESSURE
NEUTRON
SCATTERING

Stefan Klotz

CRC Press
Taylor & Francis Group
Boca Raton London New York

CRC Press is an imprint of the
Taylor & Francis Group, an **informa** business

A TAYLOR & FRANCIS BOOK

CRC Press
Taylor & Francis Group
6000 Broken Sound Parkway NW, Suite 300
Boca Raton, FL 33487-2742

First issued in paperback 2016

© 2013 by Taylor & Francis Group, LLC
CRC Press is an imprint of Taylor & Francis Group, an Informa business

No claim to original U.S. Government works

Version Date: 20120522

ISBN 13: 978-1-138-19921-7 (pbk)
ISBN 13: 978-1-4398-3562-3 (hbk)

Library of Congress Cataloging-in-Publication Data

Klotz, Stefan.
    Techniques in high pressure neutron scattering / Stefan Klotz.
       p. cm.
    Summary: "This book examines the technique of high pressure neutron scattering as a probe for the study of the atomic and magnetic structure of a wide range of materials in diverse fields. Written by a leader in the field, it provides an up-to-date review of experimental and technological aspects of high pressure neutron scattering. Suitable for researchers from various areas of the physical, chemical, and biological sciences, the text presents a useful resource for engineers and scientists who use, adapt, and improve related devices"-- Provided by publisher.
    Includes bibliographical references and index.
    ISBN 978-1-4398-3562-3 (hardback)
    1. Neutrons--Scattering. 2. High pressure (Science)--Research. 3. Materials at high pressure--Experiments. I. Title.

QC793.5.N4628K56 2012
539.7'58--dc23   2012015767

Visit the Taylor & Francis Web site at
http://www.taylorandfrancis.com

and the CRC Press Web site at
http://www.crcpress.com

Contents

Foreword

I was delighted when I learnt that Stefan Klotz was writing this book. He has an exceptional depth of knowledge and experience of high-pressure techniques, particularly as applied to neutron scattering, and it is to be greatly welcomed that he has drawn this all together and made it available.

The adventure of extending the pressure range of general neutron scattering to 10 GPa and beyond began with a chance meeting with Michel Besson in 1988. He and his group in Paris had no previous experience of neutron scattering and we, in Edinburgh, had no facilities or know-how for technical innovation on the scale required to take multi-cubic-millimetre sample volumes beyond the then longstanding limit of about 2 GPa. Within a year, a collaboration had been set up and a prototype device had been constructed for tests and subsequent development at the ISIS neutron source in the UK. The promising results of these early tests led Besson, working with his engineers Gérard Hamel and Guillaume Weill, to devise a cleverly designed low-mass press that exploited the concept of toroidal anvils. By 1990, the basics of the Paris-Edinburgh press were in place and achievable sample pressures rapidly rose to 10 GPa, and then to 20 GPa and more after the introduction of a hydraulic ram in 1991.

Stefan joined Besson's group in early 1992, and quickly became a key member through the 1990s, centrally involved with much further innovation such as the introduction of low-temperature techniques, the steady increase of the pressure limit to 30 GPa, techniques for loading liquid and gaseous samples, and the development of the new science programme thereby made possible. Early in this period, it occurred to Besson and his colleagues that, though the Paris-Edinburgh press had been designed specifically for neutron scattering, it could readily be adapted to other uses, including high-energy diffraction on synchrotron sources. This and various developments made in other contexts led to innovations that also benefited the neutron scattering applications, such as studies at high pressure combined with high temperature, studies of liquid and amorphous samples, and opening up of inelastic scattering in the 10 GPa range, in all of which Stefan has been closely involved.

A major part of the wider applications has burgeoned in the past decade, since the untimely death of Michel Besson in 2001. In this period, Stefan has taken over leadership of the Paris group and made this rich source of world-class innovation his own. After his 20 years at the centre of this field, there can be no one better placed to provide an authoritative and comprehensive account of state-of-the-art techniques for high-pressure neutron scattering, including not just Paris-Edinburgh devices but all the major contemporary experimental techniques. Those who have benefited from his direct advice and assistance will know the high-level practical and intuitive skills he brings to bear on solving technical problems and advancing innovation. Throughout the book, he has included a large amount of the wealth of practical experience, working recipes and 'rules of thumb' that he has accumulated, working on his own research and with collaborators, but which has never before been gathered together and published in this accessible way. In high-pressure experimentation, it is often these 'tricks of the trade' that make all the difference between success and failure.

Techniques in High Pressure Neutron Scattering should prove far too useful in the experimenter's laboratory ever to spend much time gathering dust on a shelf!

Richard Nelmes
Chairman: Centre for Science at Extreme Conditions
The University of Edinburgh
1 June 2012

Preface

D.B. McWhan, a pioneer in high pressure neutron scattering, predicted in 1977 during the VIth AIRAPT conference that "...high pressure neutron scattering may be one of the more significant areas of high pressure research during the next decade". Today, four decades later, most of my colleagues in the high pressure community would qualify McWhan's prediction as a stark understatement. High pressure neutron experiments have become a major tool in condensed matter research. They have allowed us to solve numerous structures of hydrogen rich solids up to the 10 GPa range and beyond giving structural parameters with a precision close to what is possible at ambient conditions. They have allowed us to give deep insight into the behavior of magnetism as interatomic distances are significantly changed and have opened new fields in solid state research such as the physics related to quantum critical points. High pressure neutron scattering enabled us to observe lattice dynamics of solids under strong compression, giving results which are often taken as direct bench marks for *ab initio* calculations.

The nowadays well recognized importance of such techniques can be judged by the observation that the major neutron facilities have beamlines which are entirely dedicated to high pressure research. Such an evolution has only been made possible by considerable progress in high pressure techniques, which were obviously accompanied by major efforts in neutron instrumentation.

The aim of this book is to resume these techniques, with a strong emphasis of methods which became available since the early 1990s, techniques which allow neutron scattering well beyond 10 GPa. The book arose out of my implication in some of these developments and the encouragement of my colleagues to write up many of the experiences and details which existed so far only as "oral tradition." The book is intended to be for both newcomers and experienced high pressure scientists and engineers. It does not require prior knowledge of the subject.

As far as I am aware there is no similar book on this topic. Among the review articles on this subject one should mention Colin Carlile and D.C. Salter's classical paper of 1978[1] and a more recent review by John Parise[2]. Also, conference proceedings exist on two workshops devoted to high pressure

---

[1] High Temperatures - High Pressures, 10, pp. 1-27 (1978)
[2] Reviews in Mineralogy & Geochemistry 63, 205-231 (2006)

neutron scattering, the 1994 meeting in Dubna[3] and the 2004 workshop at ILL, Grenoble ("MPa4Neutrons"),[4] which might be interesting for the reader.

This book would not exist without Jean-Michel Besson and Richard Nelmes who introduced me into high pressure neutron scattering when I joined their collaboration which determined my career. Within this team I had the chance to work with Gérard Hamel and John Loveday whose contributions in this field cannot be underestimated.

I wish to acknowledge the help of the following colleagues who responded to my numerous queries and who have provided important information included in this book: C.L. Bull, M. Guthrie, R. Done, C. Goodway, Th. Strässle, O. Zaharko, N. Aso, T. Osakabe, K. Takemura, R.C. Liebermann, J. Molaison, Ch. Tulk, J. Frelat, M. Tucker, H. Stone, W.G. Marshall, S. Hull, Ch. Vettier, Th. Hansen, H. Fischer, H. Schober, E. Lelièvre-Berna, J.L. Laborier, C. Payre, Ph. Salmon, J. Kamarád, O. Prokhnenko, R. Sadikov, D. Kozlenko, M. Eremets, S. Besedin, I. Mirebeau, J.-M. Mignot, B. Annighöfer, F. Maignen, W. Pichl, H. Ahsbahs, H. Kohlmann, W.F. Kuhs, K. Syassen, M. Braden, R. Winter, Ch. Pfleiderer, Th. Theuerkauf, S. Ederyd, J. Philippe, L. Bove, and P. Loubeyre.

At various places I give information in the form of "recipes," observations, or guidelines, for which no adequate reference can be cited. This should be considered as practical experience I gathered over that last twenty years in collaborative work with my colleagues at my home institute, at the University of Edinburgh, the ISIS facility, the Laboratoire Léon Brillouin, the Institut Laue-Langevin, and the Paul Scherrer Institute.

I want to thank Richard Nelmes, Eddy Lelièvre-Berna, and Burkhard Annighöfer for carefully reading large parts of the manuscript and for their helpful comments which considerably improved the book.

My special thanks go to Alain Jeanne-Michaud (IMPMC, Paris) for his invaluable help in drawing or improving most of the figures shown in this book.

This book is dedicated to my parents who allowed me a higher education and to follow the way I have chosen.

Paris, in February 2012
Stefan Klotz

---

[3] "High Press. Res." 14, 1-220 (1995)
[4] J. Phys. Cond. Matter 17, pp. S2991-S3183 (2005)

Introduction

High pressure neutron scattering is in the vast majority of all cases carried out in so called "large volume devices" which allow sample volumes of typically 1 mm^3 or more. The reason for this is the well known limitation of the flux of neutrons sources which applies to both the classical reactor and the more modern spallation sources. This is because the production of neutrons is always related to a nuclear process and the number of released neutrons per volume (or mole of nuclei) cannot be increased substantially. This is in sharp contrast to photons which can be generated out of the vacuum by accelerating charged particles to produce the extremely intensive and brilliant beams found in synchrotron sources. To make the comparison more striking: The neutron flux on one of the most intensive diffractometers (D20 at the Institut Laue Langevin) is approximately equivalent to the photon flux of a 6 V bicycle light bulb shining at a distance of 1 km![5] It speaks for the unique properties of the neutron that neutron scattering has not disappeared since decades and still plays a major role in solid state research.[6]

The need for (relatively) large sample volumes has a major and well known impact on the maximum achievable pressure, as illustrated in Figure 1. Experience and some semi-impirical considerations show that the maximum achievable pressure P_{max} is approximately proportional to an inverse power of volume V, i.e. $P_{max} \propto V^{-n}$, with n $\approx$ 2-3. The reasons behind this are various, the most important factors being limited forces and material strength combined with intrinsic physical and practical limitations. To illustrate this: A pressure of 50 GPa (500 kbar) corresponds to 500 tonnes per square centimeter and requires hence a force of 3 x 500 = 1500 tonnes to maintain a 1 cm^3 sample at this pressure, without considering how the confinement can be technically achieved. Even if this can be solved, the material in immediate contact with the sample (for example the anvils) must sustain such pressures and be available in several cm sizes. Note that the strength of common construction materials is typically 10 times lower, i.e. less than 2 GPa for steel and 5 GPa for tungsten carbide. From these considerations it becomes clear that there cannot be a "smart engineering way" to produce high pressure at simultaneous large sample volumes within the usual constrains of space and budget: An increase of pressure translates eventually always to a reduction in volume.

---

[5]Calculated on the basis of 600 nm photon wavelength, 3 W light power, and a flux of D20 of 10^8 neutrons/second/cm^2.

[6]An interesting comparison on the impact of neutron scattering compared to other techniques can be found in: F. Mezei & H.A. Graf, Neutron News 15, p. 2-3 (2004).

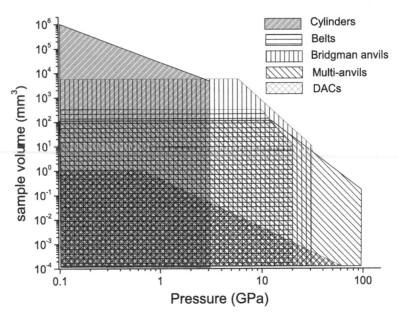

FIGURE 1
Available sample volume as a function of pressure in various high pressure devices.

Whereas this has almost no consequence for high pressure synchrotron studies where the beam size is anyway only about 10 μm, it impacts crucially the science which can be achieved by high pressure neutron scattering. Whereas almost any kind of neutron study can be carried out on a 100 mm^3 sample, i.e. complete structure determination, magnetism, and excitations, a sample volume of 0.1 mm^3 allows at the best the measurement of lattice parameters.

Is high pressure neutron scattering therefore a lost case, condemned to science in the kbar range? The answer is no, as the vivid activity in this field demonstrates, but the technical solutions are diverse and have to be adapted to the pressure range and the type of measurement. This is the very purpose of this book.

There is again a notable difference with x-ray diffraction techniques, where there is a versatile and meanwhile well-established device for high pressure studies: the diamond anvil cell (DAC). The DAC covers the entire pressure range achievable today and has become the working horse for almost any high pressure investigation on synchrotron sources, even for experiments in the kbar range. High pressure neutron scattering cannot claim such a situation, the achievable pressure range (30 GPa) is covered by various and quite different techniques, simply because for a given pressure the scattered intensity has to be maximized which imposes switching the pressure technique to the one which offers the maximum amount of sample.

There is also a fundamental difference between techniques for high pressure x-ray and neutron scattering imposed by the different types of science made with these probes. Neutron scattering is very much directed towards magnetism and low-Z materials which require in most cases only moderate high pressures (typically 0-10 GPa) but simultaneous low temperatures, ideally below 4 K. High pressure synchrotron activities are, on the contrary very much dominated by the Earth sciences which have pushed developments enabling very high pressures in the Mbar range with simultaneously high temperatures, laser heating in DACs for example.

As a result of this difference in science and community, there are very few high pressure devices which are common to both types of facilities. The Paris-Edinburgh press is a rare example of it.

Large volume pressure cells have a long history and there is substantial literature on this subject. These were the standard high pressure devices before the advent of the DAC, the devices which Bridgman and others pioneered, the machines which are still extensively used for materials synthesis under high P/T conditions. But their use for high pressure neutrons scattering is not immediately obvious. The reasons are manifold: (A) Dimensions: Neutron facilities are usually crowed places where much of the available space on a beamline is covered by detectors, and where the instruments compete which each others for each square meter. High pressure neutron devices need to be cooled, which naturally restricts their dimensions, often imposed by the existing cryogenic equipment. (B) Geometry: The cells need to be compatible with the geometry imposed by the scattering geometry, with wide neutron "windows" through the pressure cell to access the sample. (C) Materials: Neutrons have specific properties which makes the correct choice of the construction material crucial. Small neutron absorption and incoherent scattering, low temperature compatibility, and absence of neutron induced activation are of prime importance.

High pressure neutron scattering techniques is hence large-volume pressure technology with considerable constrains. This book describes the various solutions which have emerged during the past 3 decades, spanning 2-3 orders of magnitude in pressure. The philosophy of this book is to discuss various engineering concepts and to give typical examples to illustrate these. The examples are existing high pressure devices which have demonstrated their capacity, devices which are in frequent use at neutron facilities and which have produced scientific results. The book is certainly incomplete since it cannot cover many other ingenious concepts and high pressure cells - due to space restrictions - and I apologize in advance for this.

The book is organized as following. It starts with an introduction to basic engineering concepts to the elastic (and plastic) behavior of cylindrical pressure devices (chapter 1). In chapters 2, 3, and 4, construction materials are discussed, with an emphasis on mechanical and neutronic properties. The following chapters then describe various high pressure techniques, starting with liquid/gas and clamp cells (chapter 5), as well as the McWhan cell (chapter 6). Various opposed anvil techniques based on gem anvils are presented in chapter 7, followed by special designs using sapphire as neutron windows (chapter 8), as well as a chapter on uni-axial pressure cells for single crystal studies (chapter 9). The following three chapters (10-12) are devoted to Paris-Edinburgh devices, chapter (13) to high pressure metrology, and the final chapter (14) to science applications. Specific topics, tables and conversion factors can be found in Appendices A-I.

A brief word on units: despite the fact that the Pascal (Pa) is the internationally recommended unit for pressure, I have used also occasionally "kbars" when it was appropriate in the context. The same holds for forces, instead of Newton (N) I have frequently used metric tonnes (tonne), which is precisely 9806.65 N, but which for simplicity can be approximated as 10 kN.

1

Basic concepts

CONTENTS

1.1 Basic elements of material strength

High pressure technology is mechanical engineering using construction materials under particular sever conditions. It is therefore important to understand a number of frequently used engineering notions and definitions to be able to understand the performance of high pressure devices of any kind, and in particular devices used for high pressure neutron scattering. The discussion in this chapter will be important for many issues following.

1.1.1 Strain, stress, yield

Figure 1.1 shows the typical behavior of a metallic rod of length ℓ under traction. In such measurements the relative elongation $\epsilon = \delta\ell/\ell$ is determined as a function of the stress σ (more precisely engineering stress), i.e. the applied force F divided by the cross section A of the rod. For small forces (strains) the rod reacts like a spring, i.e. $\sigma \propto \epsilon$. The behavior in this region is completely elastic, which means that the rod returns to its initial length after release of the force. The constant of proportionality is called *Young's modulus* or *Elastic modulus* E, its unit is GPa.

$$\sigma = E\,\epsilon \tag{1.1}$$

1

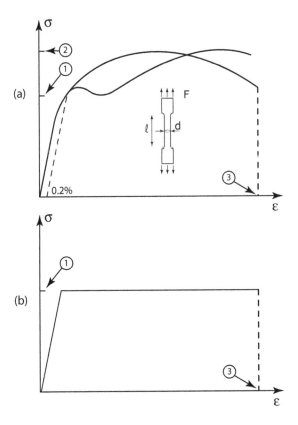

FIGURE 1.1
Stress-strain relation for a typical metal, derived from traction experiments
as sketched in the inset. Lines in (a) correspond to two different materials,
the line in (b) to an idealized elastic-plastic behavior. (1) is the yield strength
σ_Y, (2) the tensile strength σ_T, (3) the rupture elongation.

Young's moduli for steels are typically 200 GPa, for aluminum alloys 70
GPa, for tungsten carbide 280 GPa.

As the rod elongates under load, its diameter d slightly decreases. This
effect is easily observed if the experiment is carried out on a rubber rod,
for example. The negative ratio between the two relative changes is called
Poisson's ratio ν:

$$\nu = -\frac{\delta d/d}{\delta \ell/\ell} \tag{1.2}$$

Poisson's ratios for metals are typically 0.3 and do not vary substantially
across the alloys. It is easy to show that an incompressible solid has a Pois-

son's ratio of exactly 0.5. The two quantities, Young's modulus and Poisson's number completely determine the elastic behavior of an isotropic medium.

When the force on the rod and hence the elongation is increased beyond a certain point, the stress versus strain relation starts to deviate from being a straight line. This signals plastic deformation, i.e. when the load is released the rod will be slightly longer than before the traction experiment. The stress which produces a 0.2% permanent elongation is defined as the yield stress, or *yield strength* σ_Y. It is the most important quantity expressing the strength of a material. When the load on the test rod is further increased the deviation of a linear behavior becomes more and more accentuated, and various behaviors can be observed, depending on the metal and its history. In the simplest case (Figure 1.1 (a)) the stress reaches a maximum, usually far above $\epsilon = 0.2\%$, and then fails suddenly. For certain metals the stress falls slightly and then increases again. This effect is due to work hardening. The maximum stress a material can withstand is called the *tensile stress* or *ultimate tensile stress (UTS)* σ_T. Analytical calculations often assume an idealized behavior as indicated by in Figure 1.1 (b). In this case the material deforms plastically under strictly constant stress which is taken to be equal the yield stress σ_Y. The formula derived for autofrettaged cylinders discussed in section 1.2 will use such an assumption. Elongations at the point of fracture are typically between 10 and 20% for steel and other alloys. *Ductility* expresses the degree a material can deform plastically until it breaks. A metal is called *ductile* if the elongation is large and *brittle* if it is small. Certain metals and alloys can undergo a ductile-to-brittle transition as a function of temperature, i.e. they can embrittle at low temperatures. The same can occur under the influence of hydrogen.

The stress-strain behavior shown in Figure 1.1 is typical for metals and alloys. Sintered materials such as cemented tungsten carbide or sintered oxides are characterized by a very low yield strength followed by immediate rupture, i.e. these materials cannot sustain large traction. In other words these materials are brittle. This is due to the fact that fracture under tension is determined by the grain boundaries which are weak. The yield strength in these materials is difficult to determine and in any case an unreliable quantity since it varies strongly with the specimen. However, these materials can in general be strongly solicited under compression, and a more useful quantity is the *compression strength*. Compression strengths are measured with devices very similar, if not identical to those used for traction measurements. To avoid bending under compression the test rods are laterally clamped. A related quantity often cited for sintered and cemented materials such as tungsten carbide or sintered diamond is the *transverse rupture strength*. It is the stress calculated from the flexion formula required to break a specimen. A typical arrangement to measure this quantity is a beam supported near the ends with the load applied midway between the supports.

1.1.2 Hardness

Hardness is a rather ill-defined (though useful) property which characterizes the resistance of a material against deformation. Almost all types of industrial hardness measurements as discussed below are carried out by an indentation method, i.e. an indentor made of a hard material is placed on a flat test specimen and a certain force is applied for a defined duration, usually a few seconds. What is measured is the degree of penetration, by either measuring the depth or the diameter of the indentation. The different methods vary essentially by the shape of the indentor and the force applied to it. The interest of such measurements relates to the fact that for many materials, in particular metals, this quantity can be converted to the tensile strength, which provides an easy and non-destructive way to determine the strength of metallic objects. Most frequently used definitions of hardness are:

Rockwell Hardness

In this method the penetrator is either a steel sphere of specified diameter or a diamond tip with a full cone angle of 120° and a 0.2 mm tip radius. The type of indentor and the applied load define the type of scale (A,B,C, etc.). For soft materials such as copper or aluminium alloys a 1/16" steel ball is used with a load of 100 kg and the hardness is read on the "B" scale. For harder materials such as steels, a diamond cone is used with a 150 kg load and the hardness is read on the "C" scale. An example is 52HRC which means that the material has a Rockwell hardness of 52 on the C scale. What is measured is the displacement of the tip as it penetrates under three successive loads. First, a minor load of 10 kg is applied which results in a displacement of e_0 and holds the indentor in place. Then the load is increased by a major load which gives a displacement of e_1. Finally the load is decreased with the minor load still applied and the displacement e_2 is recorded. The most relevant scale for high pressure construction materials is the Rockwell C scale. It is determined through the difference $e_3 = e_2 - e_0$ by:

$$HRC = (100 - e) \tag{1.3}$$

where $e = e_3$ in $mm/0.002$.

A similar formula exists for the other Rockwell scales. The advantage of the Rockwell method is that it derives the hardness values solely from the displacements of the indentor, and not through a visual measurement of the indentation which would need the use of a microscope, such as in the Brinell, Vickers, and Knoop methods. Rockwell hardness measurements are hence easy to be carried out more or less fully automatically. HRC values are typically between 20 and 65.

Brinell Hardness

In this case the penetrator is a steel or tungsten carbide ball of diameter D (normally 10 mm) which is pushed by a constant force onto the flat surface of the material to be tested. The Brinell hardness is then determined by:

$$HB = \frac{F}{S} \tag{1.4}$$

where $F = 9.81\ k\ D^2$, k being a dimensionless number which depends on the type of material ($k = 30$ for steel, for example), and S the surface of the spherical indentation which can most easily be determined by the measured diameter of the indentor impression.

Vickers Hardness

In this case a pyramidal diamond indentor with an angle of 136° is used and a load is applied onto it for either 10 or 15 seconds. The Vickers hardness is then determined from the force F (in N) and the measured length of the diagonal d (in mmm) of the indentation by

$$HV = \frac{0.189F}{d^2} \tag{1.5}$$

Although the unit is formally Pa, usually only Vickers hardness "numbers" are cited, i.e., the numerical value. The full notation of Vickers hardness numbers is xxxHVyy/zz (example 180HV30/20), where xxx is the Vickers hardness number, yy the force on the indentor expressed in kg, and zz the duration of the indentation if it is different from 10 or 15 seconds. Typical HV values for steels are 200-600.

Knoop Hardness

In this method the indentor is a pyramid-shaped diamond with apical angles of 130° and 172.5° which produces a rhombohedral indentation with one diagonal 7 times larger than the other. What is measured using a microscope is the length of the long diagonal from which the area of impression S can be calculated. The Knoop hardness values are then calculated by

$$HK = 14.229\frac{F}{S} \tag{1.6}$$

where F is measured in kilogram-force and S in mm². The unit of Knoop hardness is hence Pa, similar to Vickers and Brinell numbers. The Knoop measurement is a micro-indentation method frequently used for ceramics and sintered material using a load of typically 1 kg. The size of the indentations are hence small, typically 0.01-0.1 mm, and a microscope is needed to carry out the measurements.

Rockwell, Brinell, and Vickers hardness values can converted to each other, over certain hardness ranges. A useful rule of thumb is:

$$HB \approx 0.95HV \qquad\qquad (1.7)$$

Another useful and important rule of thumb relates the Vickers hardness to values of the tensile strength σ_T expressed in kg/mm^2 (Figure 1.1):

$$\sigma_T \approx HV/3 \qquad\qquad (1.8)$$

As an example, a steel with a Vickers hardness of 500 has a tensile strength of approximately 167 kg/mm^2 (1.64 GPa).

For a precise conversion see the relevant tables.

Mohs Hardness

A historical and still widely used measure for hardness of minerals is the Mohs scale which spans from 1 to 10. The hardness of diamond is defined to be 10, that of talc 1, and that of 8 other minerals are given intermediate values. The hardness of a mineral is then determined by a scratch test using these gauge materials.

1.1.3   Toughness

The area underneath the curves shown in Figure 1.1 is proportional to the energy which the rod can absorb before breaking. This quantity is a measure for *toughness*. Materials with high toughness show hence elevated tensile strength and at the same time can be strongly deformed without breaking. In practice toughness is not determined from traction measurements but from so-called "notch-toughness-tests". One of them is the "Charpy test" where a pendulum of calibrated weight and velocity strikes a rod of the material to be tested containing a precut notch. Such tests are easily and quickly carried out on an industrial scale. What is measured is the energy absorbed during the fracture divided by the cross section of the rod. This quantity is then cited as "toughness" or "resilience" KCU or KCV, where the letters U and V designate the shape of the notch, its unit is J/cm^2. When comparing toughness data one should be aware of the fact that the numbers depend on the standard which might differ between countries.

Another related quantity is *fracture toughness* which measures the ability of a material to withstand the propagation of cracks under load. During plastic deformation cracks and imperfections develop in the material which eventually will lead to fracture, i.e. materials with high fracture toughness are able to resist better to the propagation of preexisting cracks and faults. Fracture toughness is an important engineering property since the occurrence of flaws is not completely avoidable in the fabrication and the use of a material. Since this ability provides an important safeguard for materials which work close to the yield limit, fracture toughness is an important property of high pressure equipment. It can be defined in terms of the *stress intensity factor* K_{Ic}:

$$K_{Ic} = \sigma \sqrt{\pi \, \alpha \, c} \qquad (1.9)$$

where σ is the critical stress required for spontaneous crack growth, c the depth of the surface flaw, and α a dimensionless number which is ≈ 1 for macroscopic samples. The unit of K_{Ic} is hence $Pam^{1/2}$. The "I" in K_{Ic} indicates crack opening according to mode I, i.e. normal stress perpendicular to the crack in a material with large thickness. Modes II and III relate to propagation under sheer stress and are much less commonly cited. Note that since K_{Ic} is a material property, equation 1.9 can be used to determine the maximum size of surface flaws allowed to avoid fracture under a given stress σ. To give an order of dimensions, $K_{Ic} \approx 50 - 200$ for steels, 10-30 for tungsten carbide, and 1-10 for ceramics. For steels and other alloys there is a clear anti-correlation between yield strength σ_Y and stress intensity factor K_{Ic}: the higher σ_Y, the lower is K_{Ic}. Steels with $\sigma_Y \approx 1.3\ GPa$ can have K_{Ic} values of up to 200 $Pam^{1/2}$, whereas for high tensile steels with $\sigma_Y \approx 2\ GPa$, K_{Ic} falls to values below 50 $Pam^{1/2}$. Stress intensity factors can be determined by methods which directly relate to formula 1.9, i.e., by using test specimens with a predefined crack of length c which are stressed until rupture occurs. These are the "compact tension" (CT) and "single edge notch bend" (SENB) geometries. For hard materials such as ceramics and gem stones which are available only in restricted dimensions, K_{Ic} values are usually obtained by an indentation test, such as used in the Vicker's hardness measurement (section 1.1.2) and applying an empirical formula such as [1]:

$$K_{Ic} = 16(H_V/E)^{1/2}F/c^{3/2} \qquad (1.10)$$

In this case H_V is the Vickers hardness, E Young's modulus, F the force applied during the indentation, and c the length of the crack which developed in the specimen.

1.2 The cylindrical pressure vessel

The following section gives a detailed discussion of the elastic and plastic properties of cylindrical pressure vessels. This is instructive and useful since the cylinder is the most common high pressure devices and one of the few mechanical problems which can be solved analytically. The results given in this section are important for understanding high pressure devices in general since they often contain cylindrical elements under load. Examples are gas and piston/cylinder cells, fretted backing seats, as well as hydraulic units.

1.2.1   The monobloc cylinder

Given a cylinder with inner and outer radius r_i and r_o, respectively, and subjected to a hydrostatic pressure p inside, the stress distributions in cylindrical coordinates is given by the Lamé equations [2, 3, 4]:

$$\sigma_\theta = p\frac{(1 + r_o{}^2/r^2)}{(K^2 - 1)} \tag{1.11}$$

$$\sigma_r = p\frac{(1 - r_o{}^2/r^2)}{(K^2 - 1)} \tag{1.12}$$

$$\sigma_z = E\epsilon_z + p\frac{2\nu(1 + r_o)}{(K^2 - 1)} \tag{1.13}$$

where $K=r_o/r_i$, the "wall ratio". It is seen from the above formula that σ_θ (as σ_r) is maximal at the bore of the cylinder ($r=r_i$), with value $\sigma_\theta=P(K^2 + 1)/(K^2 - 1)$, and falls off as $\sim 1/r^2$, see Figure 1.2. The quantity σ_θ is known as the *hoop stress*. It is by definition positive which corresponds to traction, contrary to the radial stress σ_r which is negative and which indicated compression. It is seen that the hoop stress σ_θ is always larger in magnitude than σ_r and σ_z.

To determine at which point the cylinder material starts to yield, i.e., deforms plastically, a criterion is needed which involves the three principal stresses (here σ_θ, σ_r, and σ_z) and relates them somehow to a practically accessible quantity for yield, for example the yield stress σ_Y determined from uniaxial traction measurements, see chapter 1.1. In the case of the cylinder, one might choose σ_θ as an indicator, since it is the largest of the three principal stresses:

$$\sigma_\theta = \sigma_Y \tag{1.14}$$

Equation 1.14 is called the Rankine criterion (or maximum stress criterion) for plasticity. Taking equations 1.11, this criterion predicts that yield will occur at the bore ($r = a$) at a pressure P:

$$P = \sigma_Y\frac{(K^2 - 1)}{K^2 + 1} \tag{1.15}$$

Another criterion for onset of plasticity is based on the shear stress $(\sigma_\theta - \sigma_r)$ and states that yield occurs when this quantity exceeds σ_Y:

$$(\sigma_\theta - \sigma_r) = \sigma_Y \tag{1.16}$$

This is Tresca's criterion, also called maximum shear stress criterion. With equations 1.11,1.12, and 1.13 it gives for $r = a$ (at the bore):

$$P = \sigma_Y\frac{(K^2 - 1)}{2\,K^2} \tag{1.17}$$

The nowadays most applied plasticity criterion is derived from Distortion Energy Theory and states that yield will occur when

$$\sqrt{((\sigma_\theta - \sigma_r)^2 + (\sigma_r - \sigma_z)^2 + (\sigma_\theta - \sigma_z)^2)/2} = \sigma_Y \qquad (1.18)$$

This relation is called the *Von Mises criterion* (see also Appendix A). It is believed to be the most realistic for general construction and high-tensile steels and is extensively used in Finite Element programs which calculate iso-Mises-lines to investigate where plasticity or failure is likely to occur. Iso-Mises values are calculated using a generalization of the expression on the left hand side of equation 1.18, see Appendix A.

Taking this criterium and equations 1.11, 1.12, and 1.13 it is seen that the bore ($r = a$) will yield at a pressure P:

$$P = \sigma_Y \frac{(K^2 - 1)}{\sqrt{3}\ K^2} \qquad (1.19)$$

Equations 1.15, 1.17, and 1.19 show immediately that there is an upper limit for the pressure which a monobloc cylinder can withstand, given by the limit $K \to \infty$:

$$P_{Rankin}(K \to \infty) = \sigma_Y \qquad (1.20)$$

$$P_{Tresca}(K \to \infty) = \frac{\sigma_Y}{2} \qquad (1.21)$$

$$P_{Mises}(K \to \infty) = \frac{\sigma_Y}{\sqrt{3}} = 0.577\ \sigma_Y \qquad (1.22)$$

Example: For a cylinder with K=3 made of fully hardened CuBe having a yield strength of $\sigma_Y = 1.2$ GPa, the maximal pressure which can be supported without plastic deformation is 0.62 GPa (equation 1.19). A massive increase of the wall thickness to K=10 would increase this pressure to 0.69 GPa, a meager gain of 0.07 GPa. For a high-tensile aluminum alloy with $\sigma_Y = 0.6$ GPa, formula 1.19 gives $P = 0.4$ GPa.

The above formula illustrate the fundamental problem of the monobloc cylinder: the maximal stresses occur close to the bore and an increase of the wall thickness (K) does not improve significantly its strength. For this reason, monobloc cylinders with K≥ 3 are rarely used since the considerable increase in external diameter and weight does not justify the marginal increase in strength. Since the yield strength of a high-tensile steel or alloy is at most 2 GPa equation 1.22 indicates that the maximal pressure a monobloc cylinder can sustain elastically is approximately 1.4 GPa.

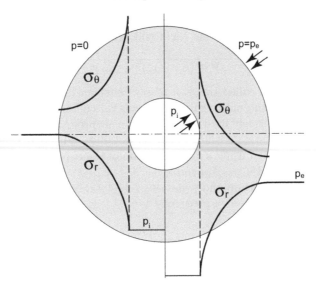

FIGURE 1.2
Stress distribution in a monobloc cylinder under two different conditions. Left:
Internal pressure $p = p_i$ and external pressure $p = 0$. Right: Internal pressure
$p = p_i$ and external pressure $p = p_e$.

1.2.2   The compound ("fretted") cylinder

One way out of the inherent pressure limitations discussed in the previous
section is the compound (or fretted) cylinder. In this case the pressure vessel
is made of two monobloc cylinders which are shrink fitted into each other,
see Fig. 1.3. Shrink fitting means that before assembling the two parts the
inner diameter of the outer cylinder (the "jacket") is slightly smaller than the
outer diameter of the inner one (the "liner"). By heating the outer cylinder it
expands and the initial dimensions are chose such that the inner cylinder then
fits into the bore of the outer. When cooled back to ambient temperatures the
inner cylinder is under radial compression from the outer cylinder. A more
popular way to achieve such radial support is to machine the outer surface
of the liner and the inner surface of the jacket with a cone of typically 1-2 $^\circ$
half angle. The dimensions are calculated such that after assembling, the liner
initially stands proud of the jacket. Pushing the liner into the jacket (with
the use of a press) then produces a similar situation as for the shrink fitted
compound cylinder. The friction between the two parts is usually sufficient to
prevent the liner to pop out after the fretting procedure.

Compound cylinders are usually made from two different materials since
using the same material has no advantages compared to "autofretted" cylin-

ders as discussed in the next section, and which are easier to manufacture. For the purpose of this discussion we will nevertheless assume that the materials are identical, or at least have similar elastic properties. In this case the contact pressure generated by an initial radial interference δ is given by:

$$P_c = \delta E / 2c \frac{[(K_j{}^2 - 1) + (K_l{}^2 - 1)]}{(K^2 - 1)} \tag{1.23}$$

where $K_j = b/c$, $K_l = c/a$, and $K = b/a$ are the wall ratios of the jacket, the liner and the compound cylinder, respectively. It can be shown that for a compound cylinder of given $K = b/a$, there is an optimal value of c in the sense that it will achieve a maximal yield strength of the vessel. The condition is:

$$c = \sqrt{ab} \tag{1.24}$$

A compound cylinder with dimensions following equation 1.24 will yield at a pressure P given by

$$P = \frac{2\sigma_Y}{\sqrt{3}} \frac{K - 1}{K} \tag{1.25}$$

The interference in this case must be:

$$\delta^{opt} = \frac{2a\sigma_Y}{E} \frac{(K - 1)}{\sqrt{3}K} \tag{1.26}$$

Equation 1.25 gives for the thick-wall limit $K \to \infty$:

$$P = 2\sigma_Y / \sqrt{3} = 1.15 \, \sigma_Y \tag{1.27}$$

Comparison with equation 1.19 for a monobloc cylinder shows that the maximum pressure can be doubled using a compound cylinder.

Example: A monoblock pressure vessel made from fully hardened CuBe ($\sigma_Y = 1.2$ GPa) has to have K larger than 4 according to equation 1.19 to withstand 0.65 GPa and remain in a fully elastic regime. Taking an inner diameter of 5 mm which is typically the sample dimension useful for neutron scattering, the outer diameter has to be larger than 20 mm. Using a compound cylinder the same performance can be achieved with $K = 1.88$ and an exterior diameter of only 9.4 mm according to equation 1.25, with c = 3.44 mm. Taking an attenuation length of 0.1 mm^{-1} (Figure 2.2), the difference in dimensions translates in an increase of scattered signal by $\propto \exp[0.1(20 - 9.4)] = 2.9$, which makes the use of a compound cylinder quite evident.

A few practical equations for compound cylinders of different materials are given in Appendix C.

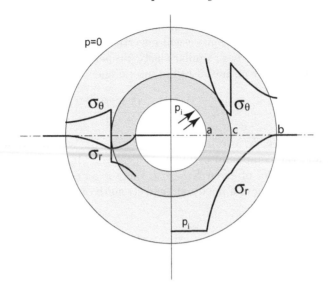

FIGURE 1.3
Stress distribution in a fretted (compound) cylinder. The left half shows the situation in the unloaded state ($p_i = p_e = 0$), the right half under internal load.

1.2.3 Autofretted cylinders

Autofrettage is another way to overcome the pressure limitations of the monobloc cylinder. It consists of starting with a monobloc vessel and applying a pressure which is beyond the yield limit given by equation 1.19, i.e. deliberately overstraining it such that the material close to the bore is plastically deformed. When the pressure is then released, the plastically deformed inner part is under compression from the outer (non-deformed) material. The plastic/elastic parts play the role of the liner/jacket discussed in the previous section for the compound cylinder. The variation of the stresses inside the cylinder wall is explained in Figure 1.3. After autofrettage, the deformations occurring during operation (experiments) remain fully elastic as long as the autofrettage pressure is not exceeded. The maximum strength of an autofretted vessel is reached when the plastic-elastic boundary is pushed up to the outer diameter, i.e. when the entire wall becomes plastic. Theoretically, the pressure needed to achieve this is given by [3, 5]:

$$P = \frac{2\sigma_Y}{\sqrt{3}} \ln K \tag{1.28}$$

This is obviously close to the burst pressure. However, overstressing a cylinder to this limit might not be desirable due to an effect called *reverse yielding*. In fact, when the pressure is released the outer parts of the cylinder will generate a compressive stress at the bore which, under certain conditions, can be larger than the strength of the material under compression. Theory shows that this can only happen for cylinders with $K > 2.2$. To ensure that such a cylinder remains in an elastic regime after the decompression phase, the elastic-plastic boundary must not be pushed beyond a value of the interface ratio $n = c/a$ given implicitly by [3]:

$$1 = \frac{K^2}{K^2 - 1} \left[\frac{K^2 - n^2}{2K^2} + \ln n \right] \qquad (1.29)$$

The pressure required for autofrettage is then given by:

$$P = \frac{2\sigma_Y}{\sqrt{3}} \left[\frac{K^2 - 1}{K^2} \right] \qquad (1.30)$$

and this is also the *maximum pressure* under which the vessel can be used subsequently in a fully elastic regime. Note that the thick-wall limit $K \to \infty$ is identical to the one for the compound cylinder, equation 1.27. To resume: For an autofretted cylinder with K<2.2 the maximum pressure under which it can be autofretted and subsequently used in a fully elastic regime is given by equation 1.28, and for a cylinder with K>2.2 by equation 1.30.

The autofrettage procedure carries certain risks and should be carried out applying a maximum of safety precautions. It is practically done by filling the bore with a solid pressure transmitting medium (for example lead) and compressing it by a piston using a press. If liquids are used the volume should be reduced to its minimum by inserting adequate spacers into the pressure chamber. The correct choice of the cylinder metal is crucial since the material must have enough resilience to support large plastic deformations. After autofrettage the cylinder is usually re-machined to correct for the slight variation in dimensions. Instead of applying pressure, the plastic deformation of the cylinder can also be achieved by machining the inner wall with a small conical angle and pushing a tungsten carbide mandrel with the same cone-angle into the bore. The procedure can be monitored by measuring the deformation of the inner or outer diameters during autofrettage. If autofrettage up to the elastic-plastic boundary n is desired, the increase of the outer diameter during the autofrettage process is given by:

$$\Delta\varnothing = \frac{2\,\sigma_Y\,n^2\,b}{\sqrt{3}\,E\,K^2}[2 - \nu] \qquad (1.31)$$

Example: A cylindrical pressure cell with $K = 4$ and a bore diameter of 10 mm made of a high-tensile aluminium alloy with σ_Y=0.50 GPa, E=72 GPa, and ν=0.33 (Table 2.5) shall be fully autofretted. Since $K > 2.2$ the elastic-plastic boundary will not be pushed up to the outer diameter but only up to the interface ratio n=1.7 according to equation 1.29, i.e. up to a radius of $c = a \cdot n = 8.5$ *mm*. This ensures that the cylinder returns to a fully elastic state after the autofrettage procedure. The pressure required to achieve this is

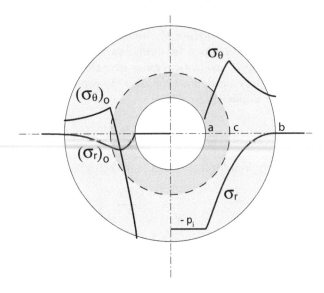

FIGURE 1.4
Stress distribution in an autofretted cylinder. The left half shows the situation in the unloaded state (denoted by "0"), the right half under internal pressure P. The elastic-plastic interface is shown as a dashed line at radius c.

$P=0.54$ GPa (equation 1.30), which leads to an increase of the outer diameter by 0.039 mm under the autofrettage pressure according to equation 1.31. The burst pressure is 0.80 GPa according to equation 1.28.

1.2.4   The segmented cylinder

The basic idea behind the segmented cylinder (Figure 1.5) is to remove the high tangential stresses appearing at the bore (equation 1.11) of the monobloc cylinder by using a sectorized inner cylinder. It is made of a material with high compressive strength, tungsten carbide for example. The forces applied by the sample on the front surface are thereby transferred on a larger surface $\propto c/a$ in contact with the containing ring. It can be shown that for a given overall $K_0 = a/b$, the maximum pressure capability of a segmented cylinder is achieved by choosing $K_s = c/a$ such that

$$K_s = \frac{1}{\sqrt{3}} K_0 \tag{1.32}$$

Note that this requires $K_0 > \sqrt{3}$. Such a cylinder can withstand a pressure P given by [3]:

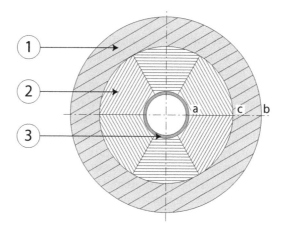

FIGURE 1.5
Segmented cylinder. (1) containing ring; (2) segments; (3) liner. (1) and (3)
are made of steel, the segments of a material with high compression strength,
usually tungsten carbide.

$$P = \frac{2}{9} \, \sigma_Y \, K_0 = 0.222 \, \sigma_Y \, K_0 \qquad (1.33)$$

and still be within the elastic limit of the containing ring.

Comparison with equations 1.19 and 1.30 shows that such a segmented
cylinder is stronger than the monobloc cylinder for $K > \sqrt{3}$, and for $K > 5$
even stronger as the autofretted cylinder. In the derivation of equation 1.33 it
was assumed that the containing ring is a simple monobloc cylinder. Obviously
it can be replaced by an autofretted cylinder in which case the performance
of the segmented cylinder is approximately doubled:

$$P = 0.415 \, \sigma_Y \, K_0 \qquad (1.34)$$

The problem with the segmented cylinder is that under load a gap between
the segments opens where the sample can extrude. This can be prevented by a
thin tube along the inner segment faces. Also the stress concentration at the tip
of the faces where the segments touch each other leads frequently to damage.
It should be noted that the above considerations explain the observation that
a pressure vessel with radially fissured (hence damaged) liner can still hold
surprisingly high pressures. An example will be given in the discussion of the
McWhan cell.

1.2.5   How strong is a cylinder?

The formulas given above indicate the maximum pressure under which a cylindrical pressure device can operate in a fully elastic regime. That does not necessarily mean that it will burst at this pressure. The reason is that the tensile stress can be considerably higher than the yield strength, i.e. the vessel can be considerably overstrained and obviously will be plastically deformed. Also the derivation of the equations is based on the assumption of an ideally plastic behavior of the material, i.e. for strains beyond the elastic limit, the stress remains constant at σ_Y as explained in Figure 1.1. Most of the construction materials will however work harden, i.e. the ultimate rupture limit in tension can be higher than σ_Y. This means that the ultimate rupture pressure of a cylinder can be somewhat higher than given by $P = 2\sigma_Y/\sqrt{3}\ln K$ (equation 1.28). This poses the important question: What is known experimentally about the bursting pressure of a cylinder? Needless to say that the answer to this question bears directly on safety issues. Systematic data on lifetime of high pressure cylinders are difficult to obtain. Sherman and Stadtmuller [5] cite sources which claim that a steel vessel with $K = 6$ ($P = 2.1$ GPa according to equation 1.28 and using a yield strength of $\sigma_Y = 1.0$ GPa) could be expected to have a lifetime of several thousand cycles up to 1 GPa, a hundred cycles up to 2 GPa, and could perhaps support 3 GPa for a few cycles. Since the number of high pressure cycles in neutron scattering measurements is usually not elevated, the use of "disposable" cylinder-pressure cells for just one experiment might be a possible mode of operation if the highest possible pressures are aimed for.

There are quite detailed studies using gas/liquid cells made of the 7000 series of aluminum alloys which are widely employed in high pressure neutron scattering. Although the conclusions are not necessarily transferable to other alloys, they are certainly useful for applications which use the same material. Figure 1.6 shows results from rupture tests carried out by Paureau and Vettier [6] (both at ambient and low temperatures), results from tests carried out at the LLB on a cell with K=3.4 described in Figure 5.1, results obtained at ISIS on a 7075-T651 cell with K=3.95 (outer and inner diameter 27.7 mm and 7.02 mm, respectively), as well as findings from a test at 300 K reported by Blaschko and Ernst [7] on a cell with K=1.9 made of an Al-alloy with a yield strength of 0.56 GPa (according to the reported composition very similar to alloys 7075). In all tests the compression medium was helium, except in the LLB rupture test where petroleum was used. The best documented test is clearly the one carried out at ISIS. The tensile and yield strengths of the source material are known from the certificates and have been measured independently to give σ_Y=0.57±0.01 GPa (hence slightly larger than indicated in Table 2.5, probably due to cold-working). For this cell, rupture occurred at 0.73 GPa after the cell has been taken once to 0.76 GPa for a short time.

Figure 1.6 resumes these results and demonstrates that the rupture pressure is reasonably well predicted by equation 1.28. In two of Paureau and

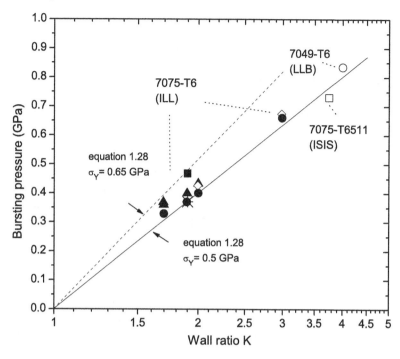

FIGURE 1.6

Experimentally determined burst pressures for cylindrical gas pressure cells made of high tensile Al-alloys as a function of wall ratio $K = \emptyset_{ext}/\emptyset_{int}$. •: T=300 K; ■: T=60 K, ▲: T=77 K, ◊: T=300 K after 20 cycles (ILL cell, Ref. [6]); ○: T=300 K, not cycled (LLB cell, F. Maignen, priv. comm.); □: T=300 K after cycling once to 0.76 GPa (ISIS cell, R. Done, priv. comm.), *: 300 K, cycled 20 times between 0.35 and 0.37 GPa [7]. Lines correspond to equation 1.28 with tensile strengths at 77 K and 300 K.

Vettier's tests the cell was cycled 20 times up to approximately half of the autofrettage pressure. Surprisingly, the cycling had no influence on the rupture pressure, i.e. there is no measurable modification of mechanical properties. This is supported by Blaschko and Ernst's tests, where the cell was also subjected to cycling close to the rupture pressure. Another interesting observation is that the rupture pressure increases with decreasing temperature which is consistent with the increase of the yield strength of Al-alloys at low temperatures, see section 2.4.2. If this is true for the 7000 aluminum alloys, it is unlikely to be valid for steels which tend to embrittle at low temperatures.

1.2.6   Summary

Figure 1.7 summarizes the pressure performance of the various types of cylinders discussed in the previous sections. There are a few remarks to be added. As said further above, the derivation of the equations is based on the assumption of an ideally plastic behavior of the material, i.e. for strains beyond the elastic limit, the stress remains constant at σ_Y (Figure 1.1). Most of the construction materials will however work harden, i.e. the rupture limit in tension is higher than σ_Y. This means that the ultimate rupture pressure of a cylinder can be somewhat higher than given by equation 1.28.

Another remark concerns temperature effects. With decreasing temperatures the yield strength of most metals and alloys *in*creases, including most materials used for neutron high pressure cells such as Al-based alloys, copper beryllium or TiZr. As a result of this, it is expected that cylindrical high pressure cells of such materials do perform better at low temperatures than expected from equation 1.28 and assuming mechanical properties quoted by the provider which refer almost always to ambient temperature. The opposite is the case for high temperatures where creep becomes an important factor, at temperatures above typically 200-300°C. Creep reduces dramatically the yield strength and the choice of the right material is crucial.

Figure 1.7 shows that the maximum pressure for elastic operation of a monobloc, shrink fitted, and autofretted cylinder rapidly saturates as the wall ratio K increases. This is not the case for the burst pressure, which increases only with $\propto ln(K)$. If a pressure cell is hence regarded as "disposable", because its cost is negligible or because the particular experiment merits to sacrifice the pressure cell, then the ultimate pressure given by equation 1.28 can be considerably high, even if materials of limited strengths are used. To give an example, a pressure cell made of the 7075 aluminium alloy with an inner diameter of 5 mm, an outer diameter of 50 mm (hence K=10) and a yield strength of 0.5 GPa can sustain pressures of 1.3 GPa, and still have a transmission of 50%. The same pressure cell made of CuBe with a yield strength of 1.1 GPa would hold up to 2.9 GPa! The transmission across 45 mm of CuBe is unfortunately only 1% but for certain samples this might be tolerable. Of course there is a high risk that such experiments will end in a catastrophic rupture. They can only be envisaged if all safety precautions are taken and if the immediate experimental environment (cryostat for example) is sufficiently shielded to avoid damage.

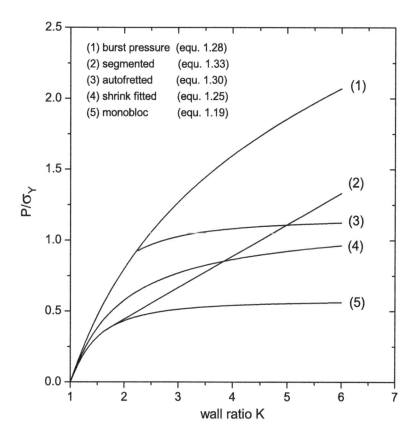

FIGURE 1.7
Limiting pressures of thick-walled vessels as a function of wall ratio K. σ_Y is
the yield strength of the vessel material.

2

Construction materials I: Nonferrous alloys

CONTENTS

2.1 Copper beryllium

Copper beryllium (CuBe) is one of the most important construction materials for high pressure devices, including cells and gaskets used for neutron scattering. The three principal reasons are: (1) Its exceptional mechanical properties which can rival with high tensile steels. After appropriate heat treatment CuBe can reach a yield strength of 1.2 GPa and a tensile strength of 1.4 GPa at 300 K. (2) CuBe is diamagnetic and hence compatible with magnetic measurements. (3) CuBe does not embrittle at low temperature, nor does it embrittle under the influence of hydrogen, in contrast to many (most) high tensile steels. This alloy is hence ideal for low temperature high pressure applications.

2.1.1 General

There are basically two types of CuBe alloys on the open market. The first type has a low Be content of up to typically 0.6 mass%. It is used for applications where high thermal conductivity is required, at the expense of mechanical performance. The second type contains Be of typically 2 mass%, and it is this material which is used for high pressure applications. The basic mechanism of hardening is due to the fact that Be has a solubility in copper of up to 2.7% at high temperatures, but only 0.2% at 300 K. When Be is dissolved in Cu at around 800°C in the α-phase (fcc) and rapidly quenched to 300 K, it forms a supersaturated solid solution. When this alloy is then heated to ∼300°C, the Be precipitates in form of a fine dispersed γ-phase. This mechanism is, as such, a quite standard example of precipitation hardening. What makes CuBe outstanding is the fact that the anneal conditions (∼300°C, in air) can be easily produced in any laboratory and that the time scale of the annealing is such that the strength can be adjusted over a wide range of typically 0.5 to 1.4 GPa.

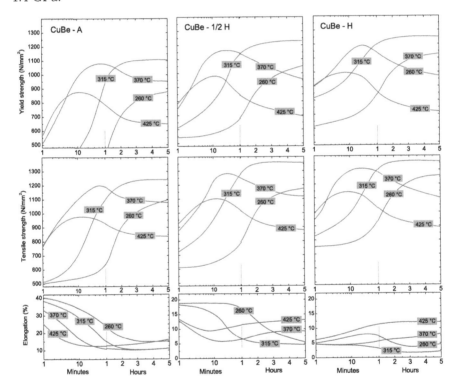

FIGURE 2.1
Mechanical properties of CuBe as a function of annealing time and temperature. Grade "A" refers to material without coldworking, "1/2H" and "H" to material which was hardened by various degrees of cold working. Source: Brush Wellman.

Figure 2.1 shows how the yield strength varies as a function of annealing time and temperature for the widely used CuBe 25 type alloy. The maximum

TABLE 2.1

Chemical composition of some high-tensile CuBe alloys available from Brush Wellman. Numbers are in mass%, the balance is Cu.

Type	Be	Co	Ni	Co+Ni	Co+Ni+Fe	Pb
25,190,290	1.80-2.00	-	-	0.2 min	0.6 max	<0.02
M25	1.80-2.00	-	-	0.2 min	0.6 max	0.2-0.6
165	1.60-1.79	-	-	0.2 min	0.6 max	-

strength is achieved at 315°C, after at least one hour, and further annealing does not change appreciably the mechanical properties. For temperatures considerably higher, the yield strength does not reach its maximum value and decreases after a few tenth of minutes.

2.1.2 Composition and nomenclature

Table 2.1 gives compositions of several CuBe grades from Brush Wellman.

Apart from the 2% Be, the commercially available alloys contain small amounts of Co, Ni, Fe, and in some cases also Si and Al. Some of these additions, such as lead, improve the machining conditions. Iron is the main source for the well known increase of the susceptibility at low temperatures [8]. Material from other sources than Brush-Wellman (such as NGK Berylco, for example) have identical Be composition but differ slightly in the relative amount of the trace metals. For neutron scattering applications these additions are irrelevant with the exception of cobalt which absorbs and activates. Alloys free of Co (and Fe) can be produced, but at considerable cost which can be hardly justified for neutron scattering applications. CuBe is commercialized in various wrought (i.e. forged or drawn) forms such as plates, bars, rods, and tubes. As a result of this, CuBe pieces have unavoidably more or less texture, and this translates to a certain extent in its neutron attenuation properties, as discussed later on. For plates and sheets, the material can be obtained cold worked which leads to an increased strength, in addition to the temper-hardening. The degree of cold working, i.e. reduction in thickness, is specified by labels such as 1/4H, 1/2H, 3/4H, and H ("quarter hardened",..., "fully hardened"). Similar to other alloys, it is desirable to use CuBe in a wrought state instead of as-cast condition. The reason is that metals in the latter condition tend to contain voids (unless vacuum casting has been used) which are eliminated in wrought material. Forging or drawing also improves the fine structure of the metal which leads to an increase in toughness.

2.1.3 Physical and mechanical properties

The most relevant alloy for high pressure application is the alloy 25. It comes under different commercial names depending on the provider, for example

TABLE 2.2

CuBe Alloy 25: some mechanical and physical properties

Yield strength	1.1 GPa (300 K); 1.5 GPa (20 K)
Young modulus	131 GPa (300 K)
Poisson's ratio	0.285 (300 K)
Ductility (elongation)	4% (300 K); 10% (20 K)
Toughness (Charpy V-notch)	7 J (20-300 K)
Thermal expansion	$17 \ 10^{-6} \ K^{-1}$
Thermal conductivity	105 W/mK
Susceptibility	$6 \ 10^{-8} \ m^3/kg$ (4.75 K)

Berylco 25 from NGK Berylco or ST 25 from Stainless. Table 2.2 gives some of its mechanical and physical properties, in its fully hardened state.

It is seen that the tensile strength increases considerably with decreasing temperature, reaching a value 30% higher at 20 K compared to ambient temperature. As similar behavior is observed in other alloys as discussed further below, i.e. TiZr and Al-alloys. Also, the ductility at low temperature is large, and the toughness does not degrade at low temperatures. These properties make CuBe an ideal construction material for high pressure cells operating at low temperatures.

2.1.4   Neutron absorption

Copper beryllium is an important construction material for gas/liquid cells and also used as gasket material in various opposed-anvils pressure cells. For such applications the incident and diffracted neutrons have to cross typically 5-10 mm of material which leads to considerable attenuation as a function of the wavelength. In addition, CuBe is usually purchased in wrought form, such as bars or plates, and the process of mechanical deformation leads to texture in the material. Figure 2.2 shows attenuation measurements carried out on a hardened CuBe disc of 10 mm thickness in two different orientations, once with the beam along the cylinder axis, once perpendicular to it. Visible are distorted Bragg edges with considerably different intensities due to texture. As a result of this the attenuation can vary by as much as 50%, such as at 4.2 Å.

2.2   Titanium-zirconium (TiZr)

We refer here to the alloy with null scattering composition which has considerable importance for high-pressure neutron scattering [9]. The bound coherent scattering length of Ti and Zr is -3.44 fm and +7.16 fm, respectively. Therefore,

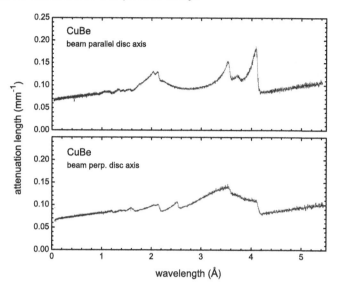

FIGURE 2.2
Attenuation of an annealed CuBe-disc as a function of wavelength, and two different orientations of the disc with respect to the beam. Source: University of Edinburgh/Paris (unpublished).

an alloy (solid solution) of Ti and Zr with a mole ratio Ti/Zr of 7.16/3.44=2.1 is completely null-scattering, i.e. an alloy with a molar composition of 67.7% Ti and 32.3% Zr will not produce Bragg reflections in a neutron beam. Since Titanium has a molar weight of 47.88 g/mole and zirconium of 91.22 g/mole, this alloy has a mass ratio of 52.5% Ti and 47.5% Zr. Fortunately, the two elements are miscible in this range of composition, and the resulting alloy has excellent mechanical properties while retaining a high neutron transparency.

2.2.1 Production

The alloy is produced in billet form applying a double vacuum-arc smelting process using nuclear grade sponges of titanium and zirconium. Residual impurities are kept to a minimum, with the hafnium (a neutron absorber) content in particular kept to below 0.004% by weight. In the as-cast form, the alloy contains large amounts of residual gas porosity, normally in the form of spherical gas pores, and it is indispensable to remove these defects. These faults are significantly reduced, if not eliminated, by subjecting the billet to a process of hot isostatic pressing (HIP). It involves the material being heated to at least 1000°C in an argon atmosphere under pressure of 1850 bar. This has the effect of reducing both the size and quantity of residual voids and produces a dense, well bonded structure, as demonstrated in Figure 2.3. Without such treatment the alloy is more or less unusable for most high pressure applications, in particular gaskets and pressure cells. Vacuum leak tests carried out at ISIS on a specimen of 0.5 mm thickness to pressures down to 10^{-5} mbar show no evidence of leak paths.

FIGURE 2.3
Surface of as-cast TiZr (left, magnification 50x) and after hot isostatic pressing (HIP) treatment (right, magnification 400x). Source: R. Done, internal report, ISIS.

2.2.2   Some physical properties and hazards

TiZr is at ambient conditions in the α-phase which has hexagonal close-packed structure. It transforms at 650°C into the body centered cubic (Im3m) β-phase which melts at 1550°C.

    TiZr is non-toxic and safe to handle in solid form. There is however a considerable fire hazard associated with the processing and use of TiZr. The metal has a greater heat of combustion than most other metals and may ignite easily when in finely divided form. This applies particularly to foils, turnings, and grindings. Once ignited, the alloy will burn fiercely, giving off intense heat and is difficult to extinguish. Water reacts with burning TiZr to give hydrogen. The alloy will burn in an atmosphere of carbon dioxide or nitrogen. Generally speaking, the maximum service temperature should be held below 200°C above which the alloy reacts with gases which leads to risk of self ignition. Above 200°C zirconium and probably also TiZr reacts with halogen gases, fluorine chlorine, bromine, iodine, and halo-carbons, including carbon tetrachloride, Freon, and Teflon. Nitryl fluoride (FNO_2) reacts with zirconium metal at room temperature to produce a glowing white incandescence.

TABLE 2.3

Properties of null-scattering TiZr. Source: ISIS, internal report.

Space group:	P6₃/mmc (194, hcp)				
Lattice param. (Å)	a=3.051	c=4.859			
Positions	atom	x	y	z	occ
	Ti	1/3	2/3	1/4	0.68
	Zr	1/3	2/3	1/4	0.32
Density:	5.23 g/cm³				
Young's mod. (GPa):	90-95 at 300 K				
	110 at 4 K				
	102 at 77 K				
Yield strength (MPa)	545-700 at 300 K				
	1320 at 4 K				
	920 at 77 K				
	545 at 473 K				
	420 at 773 K				
Tensile strength (GPa)	700-840 at 300 K				
	1320 at 4 K				
	1090 at 77 K				
	690 at 473 K				
	540 at 773 K				
Elongation:	typically 8% at 300 K				
Bulk modulus:	148 GPa				

TiZr can be machined on standard lathes and milling machines. The material has machining properties similar to that of stainless steel. Sharp cutting tools should be employed to reduce the amount of friction and a continuous supply of coolant must be provided throughout the machining operation to prevent overheating. All swarf must be removed frequently to avoid the accumulation on the machine. Remove any nearby combustible material. Workshops must be equipped with appropriate fire extinguishing equipment. Water, foam, chemical liquid, gas, or dry powder extinguisher should not be used.

2.2.3 Behavior under pressure

TiZr is used as gasket material in opposed-anvil neutron cells (with toroidal anvils in Paris-Edinburgh pressure cells, for example), and it is therefore important to know its behavior in the multi-GPa range. At 300 K, pure Ti and Zr transform both from the ambient pressure hcp α-phase to the hexagonal ω-phase (P6₃/mmm, Z=3) at respectively 5 and 7 GPa [10, 11, 12], with a volume change of 1-2% [13, 10]. At higher pressures the ω phase transforms to the bcc β-structure at 30 GPa in Zr, and at 40 GPa in Ti, though the latter seems to depend on the particular pressure-temperature path. The volume change in both cases is again 1-2%. For the null-scattering $Ti_{0.67}Zr_{0.33}$ the transformation sequence is unknown, but diffraction data exist for the equiatomic TiZr alloy. This alloy shows at 300 K a sluggish transition from the α- to the ω-phase which starts at 10 GPa and appears to extend to 25 GPa, accompanied by a volume change of 2% [13]. Further compression leads to the ω-phase at 50 GPa with a small volume change of 0.3% [13]. The same diffraction data indicate that the stability domain of the β-phase decreases with increasing temperature which leads to a α-ω-β-triple point at approximately 8 GPa/820 K [14]. One might hence conclude that given the observations in the pure metals as well as the equiatomic TiZr alloy, it is very likely that the null-scattering alloy used for neutron applications shows a very similar transition sequence. Interestingly null scattering TiZr is used as gasket material up to 30 GPa without any known indication of a reduction of its pressure performance. The phase transitions which are likely to exist in this alloy in this pressure range seem to have little influence on the mechanical properties, probably since the transitions are sluggish and extend over several GPa.

2.2.4   Mechanical properties

At 300 K, TiZr has a yield and tensile strength comparable to that of stainless steel, i.e. approximately 600 and 750 MPa, respectively. The values given in Table 2.3 were obtained from measurements at ISIS on a total of 13 ingots from three different companies delivered over the last 20 years. In the first material purchased in the early 1990s from a British provider (IMI Titanium Ltd, Birmingham), test specimens were investigated as a function of temperature at 4 K, 77 K, 300 K, 473 K (200°C), and 773 K (500°C), and these results are reported in Table 2.3 and Figure 2.4. Although the numbers at low and high temperatures are probably not too reliable they certainly indicate that the mechanical properties are strongly temperature dependent. The yield and tensile strengths seem to increase by almost a factor 2 from room temperature to 4 K. These findings are supported by the observation that TiZr gaskets used in Paris-Edinburgh type cells are less deformable at low temperatures, and also less ductile as they crack more easily under strong deformation. Figure 2.4 compares these early data with later 300 K measurements using material from a different source, together with estimated extrapolations to high and low temperatures. There is a notable scattering in these data which is thought to come from different levels of oxygen impurities which are typically in the order of 0.1%. Variations in the conditions of the HIP process might add to the scatter in the data. A more recent measurement on two samples from a 50 kg ingot gave values of 690/686 MPa and 840/833 MPa for the yield and tensile strength, respectively, and a Young's modulus of 93.8/99.3 GPa. Some of the differences might not reflect real material differences but differences in the test procedures. A report on material provided for LLB in 1988 from a French source (UGITEC) cites numbers which are considerably higher (yield strength of 762 MPa and a tensile strength of 923 MPa) and appear to be unrealistic. To resume, the range cited in Table 2.3 for 300 K are the currently accepted values which reflect the scattering over specimens from the 13 different ingots at ISIS, the large majority provided by the British company New Metals Chemicals. Poisson's ratio is apparently not known, but pure Ti and Zr have values of 0.32 and 0.34, respectively.

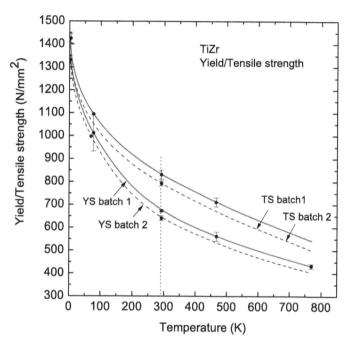

FIGURE 2.4
Yield and tensile strength of TiZr as a function of temperature determined on material of two different batches from the same provider, see text for details. Symbols correspond to measured data, lines to interpolations. Source: I. Bailey, ISIS (internal report).

2.2.5 Neutron related properties

Although perfect null-scattering TiZr does not produce Bragg reflections, its incoherent scattering shows nevertheless some structure in its Q-dependence which is important to consider if the sample signal is week. Figure 2.5 shows a diffraction pattern of a TiZr can used as sample container, obtained at the D4 diffractometer at the ILL. What is observed are well defined oscillations over the entire Q-range, as well as a few weak Bragg reflections. Simply speaking, the oscillations are due to the short range species-species order, i.e. the fact that a Ti atom is preferably surrounded by a Zr atom, and vice versa. The Bragg reflections witness obviously of a non-perfect stoichiometry, most probably on a very local scale, i.e. some degree of non-homogeneity. X-ray

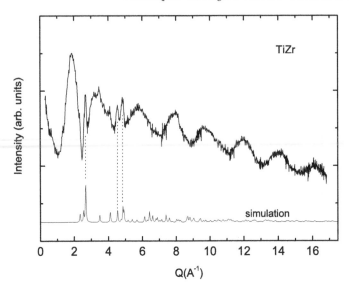

FIGURE 2.5
Diffraction pattern taken of a TiZr can (external diameter 6.5 mm, internal diameter 5 mm, illuminated length 28 mm), taken at D4 (ILL), wavelength 0.497 Å. Below: Simulated pattern with a slightly off-ideal molar composition of 65% Ti and 34% Zr. Dotted lines mark visible Bragg reflections (101), (103), and (112)/(201). Source: H. Fischer, ILL, private communication.

diffraction shows that TiZr is textured as a result of its production process (W. Crichton, priv. comm.), in particular the high isostatic pressing procedure. For this reason the relative intensities of the residual Bragg reflections varies during an experiment where the material is strongly deformed, such as in the case of TiZr used as gasket material. The diffraction pattern shown in Figure 2.5 is very similar to patterns obtained from TiZr of other sources, and used for gasket material, for example. Note that for most diffraction measurements on crystalline samples these details are irrelevant since the signal of a typical sample is several orders of magnitude larger than the amplitudes of the oscillations produced by TiZr. However, for measurements on disordered materials under high pressure the knowledge of the background produced by TiZr is crucial and essential for the data analysis.

Figure 2.6 shows measurements of the attenuation of null-scattering TiZr. As a rule of thumb, at a wavelength of 2 Å, a neutron beam looses approximately 5% of its intensity across 1 mm of TiZr.

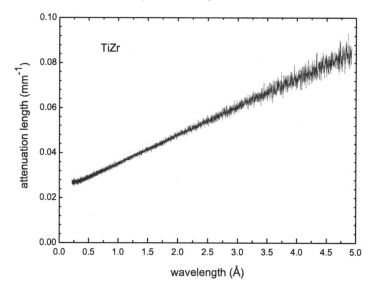

FIGURE 2.6
Wavelength dependence of attenuation of null-scattering TiZr. Source: University of Edinburgh/Paris (unpublished).

2.2.6 Hydrogen compatibility

TiZr is clearly incompatible with hydrogen or deuterium, in particular at high temperatures, as reported by E. Gray (Griffith University, internal report). In these studies a bar of 3 mm diameter was exposed to pressurized deuterium and the gas uptake was observed at various temperatures. At 200°C and 10 bar deuterium was readily absorbed by TiZr up to a deuterium-to-metal ratio of 1.57. When the sample was depressurized under argon atmosphere it completely decrepitated to a fine powder. The experiment was repeated at 150°C which gave a deuterium-to-metal ratio of 0.5 after three days of exposure, and again the sample partially decrepitated upon pressure release. At 100°C no uptake was observed until the deuterium pressure reached 600 bar where the rod started to absorb the gas for the following 12 days to reach a deuterium-to-metal ratio of 1.89. The same experiment carried out at 50°C and 700 bar gave no indication of gas absorption. The conclusion of this is that TiZr appears to be safe in contact with deuterium (and hydrogen) at pressures of several hundred bars if the temperature stays below 50°C, but it is clearly incompatible with these gases at temperatures above. This strongly limits its use for pressure cells and gaskets which are in contact with these gases. Given the fact that hydrogen and hydrogen rich samples play a major role in research carried out at neutron facilities, a better understanding of

TABLE 2.4
Typical mechanical properties of high tensile Ti-alloys,
quenched and annealed. Units of σ_Y (yield strength) and
σ_T (tensile strength): GPa. A: elongation.

short name	DIN designation	σ_Y	σ_T	A
TA6V	TiAl6V4ELI	1.05	1.15	12 %
TA6V6E	TiAl6VSn2	1.15	1.25	10 %

the influence of hydrogen on construction materials, in particular metals and
alloys, appears urgent.

2.3 High tensile titanium alloys

Some titanium alloys have strengths comparable to CuBe. The best known
are the materials with designation TA6V (TiAl6V4ELI) and TA6V6E
(TiAl6VSn2) which contain Al and V as the main alloy elements. Table 2.4 resumes some of their mechanical properties. Their application for neutron scattering devices is not very common compared to CuBe and the null-scattering
TiZr alloy. Compared to CuBe these alloys have lower attenuation (despite
the larger absorption cross section), and contrary to TiZr they are readily
available in large dimensions. Ti-alloys are non-magnetic (paramagnetic) and
find hence applications for pressure cells which are exposed to high magnetic
fields, see Figure 9.2 for an example.

2.4 High tensile aluminium alloys

Aluminium alloys play an important role as construction material for neutron
applications due to the very low neutron absorption and incoherent cross
section of Al. One millimeter of pure aluminum attenuates a thermal neutron
beam by approximately 1%. Aluminium or aluminium alloys would hence be
an ideal construction material for high pressure neutron cells. Unfortunately
even the strongest Al-alloys have yield and tensile strengths which are two
to three times lower than those of high tensile steels. Nevertheless certain
Al-alloys are widely used in gas- and liquid cells up to 0.8 GPa, as will be
discussed in this chapter.

2.4.1   General

Aluminum alloys are designated by a 4-digit number where the first digit indicates the major alloy element. The 1000 series is almost pure Al (>99%). For the 2000, 3000, 4000, 5000, 6000, and 7000 series the principal alloying elements are copper, manganese, silicon, magnesium, magnesium+silicon, and zinc, respectively. These alloys can be classified into two broad categories: non-heat-treatable and heat-treatable alloys. For the non-heat-treatable alloys the improvement of the mechanical properties relies on solid-solution hardening and work-hardening. The 1000, 3000, 4000, and 5000 series belong to this class. In the heat treatable alloys the principal strengthening effect is derived from the precipitation of a secondary phase as the alloy is quenched from high temperatures. The 2000, 6000, and 7000 series fall into this category. A widely used type of alloy called "Duraluminium" or "Dural" belongs to the 2000 series. The degree of work hardening and heat treatment (tempering) is indicated by the letter "H" and "T", respectively, followed by a number of up to 3 digits. Example: the degree of work hardening increases from H12 to H18. T6 indicates quenched and annealed above ambient temperature, T651 is a variant of T6.

The Al-alloys with the highest strengths are those which are heat-treatable, in particular the 7000 series, such as the 7075-T6. The higher strength levels attainable in this class are accompanied by a deterioration of some other properties such as corrosion resistance, weldability, and ductility. Similar to other alloys (such as CuBe), the mechanical properties of Al-alloys of any kind can be improved by work-hardening, i.e. even for the heat-treatable types. Laminated plates can have yield- and tensile strengths considerably higher than undeformed material. However, the mechanical properties are strongly anisotropic which is in general undesirable for high pressure applications.

2.4.2   Mechanical properties

Table 2.5 gives physical and mechanical properties of two high-tensile Al-alloys, 7075-T6 and 7049A-T6. These alloys have the highest strengths of all Al-alloys and are therefore relevant for high pressure applications. All other types, including other grades within the 7000 series, have lower yield and tensile strengths and will not be discussed here. Aluminium alloys with the composition very close to the 7075 and 7049 material are commercialized under various names, for example Alumec, produced by Alcoa, or Fortal produced by Pechiney (Alcan). The values for strengths found in the literature and quoted by the producers scatter because of different degrees of work hardening which depends on the shape and dimensions of the material. Laminated plates and rods have higher strengths than undeformed material. In plates the strength close to the surface is systematically higher than in the center.

Aluminium alloys improve their yield and tensile strengths at low temperatures without loosing their toughness, see Figure 2.7. The increase in σ_Y and

TABLE 2.5
Properties of 7075-T6 and 7049A-T6 high tensile Al-alloys. Data on
yield and tensile strength are typical for rods with $\varnothing{<}100$ mm.
Source: Euralliage, France, and MatWeb - Material Property Data.

Grade		7075	7049 A
Composition (mass%, balance: Al)	Zn	5.1-6.1	7.2-8.4
	Mg	2.1-2.9	2.1-3.1
	Cu	1.2-2.2	1.2-1.9
	Fe	0.50	0.50
	Si	0.40	0.40
	Mn	0.30	0.50
	Cr	0.18-0.28	0.05-0.25
Density (g/cm^3)		2.83	2.84
Yield stress (GPa, 300 K)		0.50	0.53
Tensile stress (GPa, 300 K)		0.56	0.61
Elongation (%, 300 K)		7	5
Toughness K$_{IC}$ (MPam$^{1/2}$)		26-34	24-31
Young modulus (GPa)		71.7	71.7
Poisson's ratio		0.33	0.33
Specific heat (J/gK)		0.86	0.96
Thermal conductivity (W/mK)		153	154
Lin. therm. exp. coeff. (10^{-5}/K, 300K)		2.3	2.3

σ_T is typically 20% or more with ductilities (elongations) remaining in the 10%
level or higher. However, the same mechanical quantities decrease strongly at
only moderate high temperatures, i.e. in the 100-150°C range. This decrease is
particularly pronounced for the 7075 alloys which are the standard materials
for high pressure neutron cells. Figure 2.7 indicates that for pressure devices
used beyond 100°C, certain alloys of the 2000 series (for example the 2024 or
the 2618) are in fact preferable.

2.4.3   Neutron attenuation

Aluminum and aluminum alloys have high neutron transparency due to the
very small absorption and incoherent cross section of Al, combined with a
relatively small coherent scattering power (Appendix G). Figure 2.8 shows
measurements of the attenuation of the 7075 alloy (Table 2.5). As a rule of
thumb, at a wavelength of 2 Å, a neutron beam looses approximately 1% of
its intensity across 1 mm of this material and alloys of similar composition.

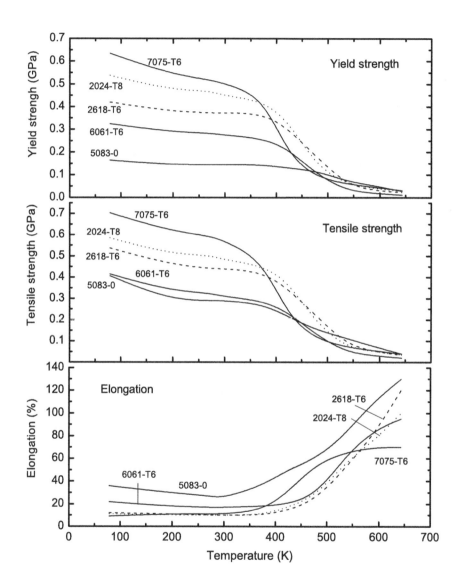

FIGURE 2.7
Mechanical properties of aluminium alloys as a function of temperature (after 10^4 hours of exposure at the indicated temperature). Source: J.R. Davies et al. (edts.), Aluminum and aluminum alloys, ASM International, 1993.

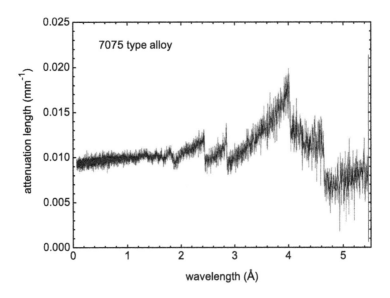

FIGURE 2.8

Wavelength dependence of attenuation of the 7075 aluminium alloy. Note the extremely low attenuation length beyond the first Bragg edge at 4.7 Å. Source: University of Edinburgh/Paris (unpublished).

3

Construction materials II: Steels and super-alloys

CONTENTS

3.1   General

Steel is an alloy of iron and various other elements, with carbon playing a major role next to nickel and chromium. The importance of steel as a construction material in high pressure devices is evident and obviously includes applications for high pressure neutron scattering. The literature on steel fills entire libraries, and there is no point to aim for a comprehensive discussion in this chapter. Instead, this chapter will give basic and general elements of the metallurgy of steels, some of their physical and mechanical properties, the classification of steels and their designations, i.e. the various systems of naming. We will focus on high-tensile steels and some special alloys since high pressure devices tend to work under high stresses. The emphasis is on steels and alloys which are used or have the potential to be used in high pressure equipment for neutron scattering.

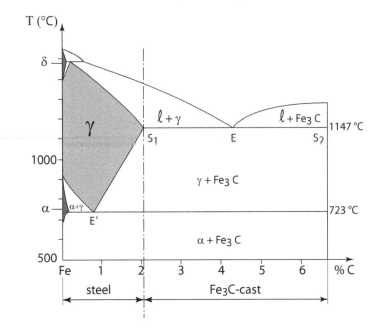

FIGURE 3.1
Iron-carbon phase diagram at ambient pressure. The carbon content is given in mass%, ℓ is the liquid, E' and E the eutectic compositions at 0.77% and 4.2% carbon. α and γ are the bcc and fcc phases, respectively.

3.2   The iron-carbon phase diagram

The simplest steels are composed of two major elements, iron and carbon. It is therefore useful to discuss the Fe-C composition phase diagram which will illustrate the basic mechanisms of hardening and help to understand the metallurgy of more complex steels which include further elements such as Ni, Cr, Mo, Si, V, and Mn.

Pure iron has at ambient pressure three phases, all of them being cubic: the ambient-temperature body-centered α-phase which is ferromagnetic, the high temperature face-centered γ-phase stable between 910°C and 1403°C, and another bcc phase (δ-phase) between 1403°C and the melting point at 1535°C. We ignore the latter since its stability field is small. Figure 3.1 shows that γ-Fe can incorporate up to 2% carbon whereas α-Fe at most 0.05%. The corresponding solid solutions (alloys) are called austenite and ferrite. If austenite is cooled slowly beyond its stability field it will expel the carbon in form of Fe_3C (cementite), or under certain conditions as pure graphite. At the end there will be an inhomogeneous material composed of α-Fe and Fe_3C. For an eutectic composition of 0.77% carbon, the resulting microstructure is called perlite. The situation is different when the austenite is cooled rapidly, i.e. quenched. In this case the carbon has no time to leave the crystal lattice and is trapped on interstitial sites. The resulting structure is tetragonal in average but locally distorted, and called martensite. Martensite is considerably harder than pure iron since the lattice distortions prevent the movements of defects which govern plasticity. As a consequence, the hardness increases with the amount of carbon. This is the basic mechanism of hardening of steels which transforms a rather soft element into a material with exceptional mechanical properties. Although martensite is hard, it is not very useful since its toughness is rather low, i.e. it breaks easily. For this reason the material is tempered to temperatures just below the point where it transform back to austenite. This procedure increases the toughness without reducing considerably its yield and tensile strengths.

3.3   Other alloy elements

Modern steels contain more elements than iron and carbon. Steels with less than typically 0.5% for each alloy element (beyond carbon) are considered as unalloyed, steels with a total alloy content up to 5% as "low alloy steels" and above as "high alloy steels." Several elements are naturally present due to the fabrication process such as oxygen, silicon, and phosphorus, others are deliberately added such as nickel and chromium. These elements shift the phase boundaries of Figure 3.1 and make the diagram obviously more complex, but in general there are still stability domains of α and γ phases. There are some elements which promote the α-phase, others the γ-form. A major effect of these additional elements is that they delay the diffusion of carbon and therefore ease the procedure of quenching: in some cases simple air cooling is enough to form martensite. There are some elements which are clear steel poisons, i.e. their presence lead to embrittlement. These are H, O, S, P, As, Sb, and Sn. For high tensile steels the maximum amount of O, S, and P should be below 0.01%. The case of hydrogen embrittlement will be discussed in more depth in section 3.8. The characteristic influence of some other elements on steels are:

1. Cobalt: Co does not form carbides, it hampers the grain growth at high temperatures and is used for high-temperature steels. For neutron applications it is disadvantageous since it absorbs neutrons and forms ^{60}Co which activates the steel.

2. Chromium: Cr strongly reduces the speed needed to produce martensite which allows quenching in oil or air. Chromium forms carbides which can be beneficial for wear resistance. High chromium

contents above 10% give corrosion resistant steels. Chromium promotes the α-phase.

3. Manganese: Mn reduces strongly the critical speed of cooling to produce martensite. Mn promotes the γ-phase and renders steels with Mn-contents larger than 12% austenitic if the carbon content is sufficiently high. Mn produces sulfides and suppresses the formation of iron sulfides which are a steel poison.

4. Molybdenum: Mo reduces strongly the critical speed of cooling to produce martensite. It promotes the growth of small grains which increases the strength. It also increases the corrosion resistance and appears often in highly alloyed Cr-steels.

5. Niobium/Tantalum: Appear almost always together since they are difficult to separate. Both are strong carbide formers and are added to stabilize chemically inert steels.

6. Nickel: Ni increases considerably the toughness. It also increases strongly the stability of the γ (fcc) phase. For more than 7% and high Cr content it forms austenitic steels at temperatures well below 300 K. Austenitic steels are non-magnetic and show high strength beyond 600°C. Nickel is the key element in maraging steels as discussed below.

7. Lead: Pb is added up to 0.2-0.5% to improved the machining properties of steels since it promotes clean cutting surfaces and small turnings.

8. Silicon: Si is naturally present in iron since the ores contain already a certain amount and the furnaces used for steel production use Si-containing minerals. Silicon reduces strongly the stability range of the γ-phase, increases the strength and wear resistance. It strongly increases the yield strength and is hence used in spring steels. It is also used in high-temperature steels to increase the heat resistance.

9. Titanium: Ti has a strong affinity to oxygen, nitrogen, sulfur, and carbon. It is hence strongly deoxidizing and forms nitrides, sulfides, and carbides.

10. Vanadium: V is a strong carbide former, reduces the grain size and improves the microstructure. It also improves the weldability, increases the Curie temperature. Vanadium carbides improve the resistance to pressurized hydrogen.

3.4 Designation (naming) of steels

There are thousands of different types of steel but unfortunately not a unique way of classifying and naming them. The reason is historical: over the centuries the major steel producing countries have developed their own standards which now coexist. As a result of this, one and the same steel with a given chemical composition has different designations in different countries, and in addition, is sold under different commercial names. To give an example, the steel with the short name X17CrNi16-2 (see explanation below) is called S 80 according to the British standard and AISI 431 under one of the American standards. It is commercialized in France by Aubert & Duval as "APX."

A most useful and internationally recognized way to name a steel is by its chemical composition (as apposed to its application and/or physical properties, for example). Such "short names" arouse out of the German DIN standard. The syntax is as following:

[*] [C-cont.x100] [1] [2] ... [n] [cont. 1]-[cont. 2]-...-[cont. n]

[*] might be "C" for carbon, "G" for cast iron. But the most relevant is "X" which means it is a highly alloyed steel where at least one of the alloy elements [1]...[n] has a content [cont-1]...[cont-n] of more than 5%. If this is the case, then the alloy contents are given simply in %. If this is not the case it is simply omitted and the alloy contents given at the end are multiplied by factors which depend on the element: Cr, Co, Mn, Ni, Si, W: factor 4; Al, Be, Cu, Mo, Nb, Pb, Ta, Ti, V, Zr: factor 10; Ce, N, P, S: factor 100; B: factor 1000. If an element appears only in minor amounts (typically below 1%), the content number might be omitted. The carbon content at the begin is always multiplied by a factor 100.

Examples:

1. 35NiCrMo16: This short name indicates a carbon content of 0.35%, a nickel content of 4%, and minor amounts of chromium and molybdenum which can be found in the corresponding data sheet (which says 1.75% Cr and 0.5% Mo).

2. X17CrNi16-2: This is a highly alloyed steel with 0.17% carbon, 16% chromium and 2% Nickel. The high Cr content reveals that it is a stainless steel.

3. X2NiCoMo18-8-5: This is a highly alloyed steel containing 0.02% carbon, 18% Ni, 8% Co, 5% Mo. The extremely small carbon content and the high Ni-content between 15 and 20% reveal this steel to be a maraging steel (see below). The British standard (BS) is S 162, the American UNS standard is K92890. It is commercialized in France under the name "Marval 18."

Another way to classify steels (and certain alloys) is by its "Material Number" ("Werkstoff-Nummer" in German). This is a unique 5-digit number which is given to most major steels and alloys and which allows one to find its composition. The above mentioned "APX" steel (X17CrNi16-2) has for example the Material Number 1.4057. The "Aviation standard nr." (WL-number) would be 1.4044. A useful aid in finding the composition and classification of steels and a comparison between international standards is the German *Stahlschlüssel* [15]. It contains data on approximately 85000 entrees of steels produced worldwide. The *Stahlschlüssel* is updated every three years.

3.5   High tensile steels

These are steels with a yield strength of typically larger than 1.2 GPa, and the older literature sources often designate them as "ultra-high tensile steels." Such materials are used for example in aeronautical applications where the weight for a given strength is an important factor (example: landing gear of aircrafts). The need to minimize the weight without reducing the performance applies also for most high pressure devices, including large volume cells for neutron scattering. As a consequence, such kind of steels find extensive applications in high pressure technology, such as load frames for McWhan cells and Paris-Edinburgh type presses.

High tensile steels can be grouped into two categories, depending on the mechanism of hardening. The first class concerns steels where the hardening process is governed by carbon, the classical hardening process as explained further above. These steels are called "heat treatable" or "alloyed" steels (German: Vergütungsstahl; French: aciers alliés), though the names are misleading since most other steels are also heat-treatable and alloyed. The second category are the more modern maraging steels. They contain almost no carbon and the hardening is produced by precipitation of intermetallic compounds made of minor alloy elements. Maraging steels are a relatively new invention, they exist since the early 1960s.

3.5.1   Heat treatable / alloyed steels

These are steels with up to 2% carbon and additional elements as discussed in paragraph 3.3. The hardening process is governed by carbon: it is dissolved at high temperatures (approximately 800°C) into the γ phase followed by quenching (French: trempe) to produce martensite. The following tempering process (French: revenu; German: Anlassen) at temperatures between typically 200°C and maximal 680°C allows then to adjust the mechanical properties to the desired values. The strength of these steels increases strongly with the carbon content: The yield strength can almost be doubled as the

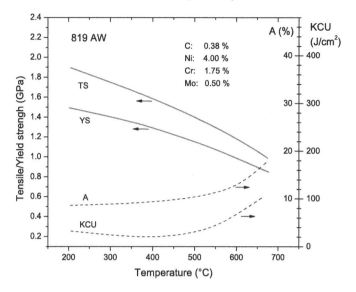

FIGURE 3.2
Mechanical properties of the high tensile steel 819 AW as a function of tempering temperature, after heating to 875°C and cooling in air. A is the elongation at 5d, KCU the Charpy toughness, see chapter 1 for definitions. Source: Aubert & Duval.

carbon content increases from typically 0.2 to 0.5%, but unavoidably at the expense of toughness [16]. The mechanism of heat treatment to achieve specific mechanical properties in this class of materials is illustrated in Figure 3.2 on the example of a steel commercialized by Aubert & Duval (France) under the trade name 819 AW. It is widely used in high pressure devices, for example in the construction of Paris-Edinburgh presses. This steel contains 0.38% carbon and only three other alloy elements: Ni (4%), Cr (1.75%), as well as Mo (0.5%), and has been remelted to remove minor contaminant elements. The heat treatment consists of quenching from 875°C to ambient temperature followed by tempering between 200°C and 660°C. Figure 3.2 shows how the strength and the toughness decreases as the tempering temperature increases, accompanied by an increase in ductility. For low temperature and aeronautic applications the steel is cooled to -80°C between quenching and tempering to remove traces of austenite.

Among the large number of heat-treatable steels we give in Table 3.1 some examples of grades which are currently used or could be relevant for high pressure devices for neutron scattering. Besides the strength, the selection criteria were the absence of Co to avoid potential activation through ^{60}Co and the availability of the material in bulk form.

TABLE 3.1

Characteristics of some high-tensile heat treatable/alloyed steels. Only Co-free alloys are listed, the base element is iron. Units: GPa (σ_Y, σ_T), % (A), J/cm^2 (KCU). Source: mostly from ref. [16].

Short name	Mat.nr	C/Si/Mn/Cr/Mo/Ni/V (%)
32CrMo12[a]	1.7361	0.32/<0.4/0.55/3.0/0.4/<0.6/0
30CrNiMo8	1.6580	0.30/<0.4/0.55/2.0/0.4/2.0/0
300 M	1.6928	0.42/1.65/0.75/0.80/0.40/1.80/0.10
30NiCrMo10	1.0773	0.30/<0.4/0.45/1.8/0.35/3.8/0
X41CrMoV-5-1	1.7783	0.41/0.90/0.30/5.0/1.3/0/0.50

Short name	σ_Y	σ_T	A	KCU
32CrMo12[a]	1.15	1.40	8	60
30CrNiMo8	1.25	1.40	10	50
300 M[b]	1.65	1.95	8	8
36NiCrMo16[c]	1.35	1.75	8	30
X41CrMoV-5-1[d]	1.6	1.9	8	30

[a]Ref. A&D: GK3; tempered at 500°C; [b]tempered at 300°C; [c]Ref. A&D: 819B, tempered at 300°C; [d]tempered at 550°C.

3.5.2   Maraging steels

The word *maraging* derives from *martensitic ageing*. These steels contain almost no carbon but a rather high amount of Ni of approximately 18% next to minor elements such as Co (5-10%), Mo (a few %), and Ti (usually up to 1.5%). When the alloy is quenched from the γ phase it forms a martensitic structure, similar to carbon containing steels. If the steel is then heated to approximately 500°C ("aged", German: auslagern), these elements precipitate in form of intermetallic compounds such as FeTi, Ni_3Mo and Fe_2Mo which causes a drastic increase in strength. The heat treatment is therefore considerably simpler than for "alloyed" steels discussed before. Also, the volume change during ageing is almost negligible so that the items can be made to final dimensions. For a given yield strength, maraging steels tend to have a higher toughness than carbon-based steels. From all these considerations it would appear that maraging steels are ideal for high pressure applications. This is only partially true. One factor is cost, the price of maraging steel can easily be twice of ordinary high-tensile steels. There is a serious inconvenience for neutron applications since the high Co-content poses a permanent risk of activation and a need for appropriate shielding. Popular types of maraging steels are grades C-200, C-250, and C-300 (Mat.nrs. 1.6359 and 1.6358), some of them commercialized under the name Vascomax. Parts of the ILL-McWhan high pressure cell are made of this steel. The equivalent grade from Aubert & Duval (France) is Marval 18 (1.6359, X2NiCoMo18-8-5).

3.6 Stainless steels

The characteristic of this type of steel is a high chromium content of at least 12%. The chromium forms with oxygen a passivation layer which renders these steels resistant to corrosion. The majority of stainless steels have only moderate strength in its bulk form which is unfortunate for high pressure applications. Nevertheless, certain types as discussed below can be hardened by appropriate heat treatment and then reach yield strengths of typically 1.2-1.4 GPa, others by extensive cold-working. Stainless steels can be classified in several categories:

1. Austenitic stainless steels. These present at least 2/3rd of all stainless steels used in everyday life. Their chromium content is relatively high (around 18%) and the large amount of Ni of 8-10% stabilizes the fcc (austenitic) structure down to ambient and even low temperatures. A typical example is X5CrNi18-10 (AISI 304). They are non-magnetic and show excellent corrosion resistance. After annealing at standard temperatures their mechanical strength is very moderate, i.e. the yield strength is typically 0.2-0.3 GPa at 300 K. However, the strength can be considerably increased by cold-working to reach values of the yield strength of > 1 GPa. High pressure tubing are often made of AISI 316 (X5CrNiMo17-12-2) or AISI 316Ti (X6CrNiMoTi17-12-2) for example. Most of the high-pressure fittings are made of this kind of material as well. A special class of austenitic steels are those which resist high temperatures, such as X6NiCrTiMoVB25-15-2 (Mat.nr. 1.4980; aviation standard no. 1.4944) and which are characterized by a very large Ni content (25%). Their yield strengths of typically 0.65 GPa at 300 K are still 0.4-0.5 GPa at 900 K. These types of stainless steels are believed to be highly resistant to hydrogen under high pressure. The membranes of hydrogen compressors are often made of this material.

2. Martensitic stainless steels. These are characterized by a chromium content of typically 13%, a low nickel content (up to 2%) and carbon content in the range 0.1-1%. The hardening process is governed by carbon similar to the process discussed for the high tensile steels. Typical examples are X20Cr13, X12Cr13 (AISI 410), and X17CrNi16-2 (AISI 431). The yield strength of these steels can reach 1.2 GPa after quenching, but drops to typically 0.5-0.7 GPa after the standard tempering at 600°C. Martensitic steels are ferromagnetic.

3. Precipitation hardenable stainless steels. These contain additional Cu, Al, Nb, or Mo on the level of a few per cent, in addition to Cr (15-18%) and Ni (typically 5-7%). Typical examples are

TABLE 3.2
Examples of stainless steels with typical strength
properties. A=austenitic, M=martensitic, F=ferritic,
PH=precipitation hardening. Units: GPa (σ_Y,σ_T), % (A,
elongation).

Designation	AISI	type	σ_Y	σ_T	A
X5CrNi18-10	304	A	0.31	0.60	30
X5CrNiMo17-12-2	316	A	0.30	0.58	55
X6CrNiMoTi17-12-2	316Ti	A	0.32	0.60	50
X12Cr13	410	M	0.31	0.52	25
X17CrNi16-2[a]	431	M	1.08	1.35	20
X6Cr13	403	F	0.22	0.50	20
X6CrMo17-1[b]	430F	F	0.50	0.80	12
X7CrNiAl17-7[c]	631	PH	1.52	1.62	6
X8CrNiMo15-7-2[c]	632	PH	1.55	1.66	6
PH 13-8 Mo[d]	XM-13	PH	1.40	1.52	10
15-5 PH[e]	XM-12	PH	1.39	1.44	10
17-4 PH[f]	630	PH	1.38	1.45	7
Custom465[g]	-	PH	1.65	1.75	12

[a]tempered at 427°C; [b]tempered at 650°C; [c]tempered at 510°C; [d]tempered
at 510°C, nr. 1.4534; [e]X5CrNiCuNb15-5-4, nr. 1.4545, tempered at
482°C; [f]X5CrNiCuNb17-4-4, nr. 1.4542, tempered at 482°C; [g]type
X2CrNiMoTi12-11, tempered at 510°C.

X3CrNiMo13-4, X8CrNiMo15-7-2 (AISI 632), and X7CrNiAl17-7
(AISI 631). The hardening process is governed by the precipitation of intermetallic compounds which involve the minor alloy elements mentioned above. These steels can either be martensitic or
austenitic, hence magnetic or non-magnetic, and their mechanical
strength can be elevated. For example the X7CrNiAl17-7 (AISI 631)
can reach yield strengths of 1.0-1.5 GPa, depending on the tempering temperature. Typical high pressure applications are stainless
steel hand pumps operating in the 0-7 kbar range where the AISI
630 (X5CrNiCuNb16-4) is used, for example.

4. Ferritic stainless steels. These are steels with typically 17% Cr and
   only carbon (below 0.1%), and possibly small amounts of Mo, as additional alloy element. They retain their bcc structure after quenching, hence they are magnetic. Typical representatives are X6Cr13
   (AISI 403) or X6CrMo17-1 (AISI 430F).

3.7 Superalloys

These are strictly speaking not steels since the major element is generally not iron but nickel or cobalt. The interesting properties of these materials are their excellent strength and resistance to creep and oxidation at high temperature up to 600-1000°C. This explains their industrial applications in turbines and rocket engines, and in high pressure research for high-temperature pressure cells. Superalloys are either nickel, cobalt, or iron based alloys and are generally of austenitic (fcc) structure. The strengthening mechanism is based on precipitation of intermetallic compounds and carbides. Most superalloys are Ni-based with a Ni-content larger than 50% and additional elements such as Co, Cr, Mo, W, Al, Ti, and Ta. Similar to stainless steels, the corrosion resistance derives mainly from chromium. The current practice is to designate superalloys by their producer rather than by a short name, for example Inconel 718 by Inco Ltd. or René 41 by Haynes. Unfortunately, even the nickel-based superalloys contain in most cases non-negligible amounts of cobalt and are hence unlikely to be useful for neutron scattering applications. These include for example the well known grades René 41 (10-12% Co), Waspalloy (12-15% Co), Nimonic (up to 20% Co), and MP35N (35% Co; this alloy is not a superalloy according to the definition given above but has outstanding strength and corrosion resistance up to 300°C). If one adds the requirement of exceptional high strength of the material in form of bars and rods there are only a few options such as the Inconel 718 or PER706, the latter being produced by Aubert & Duval (France). Both are non-magnetic. Table 3.3 gives their typical chemical composition and mechanical properties at high temperatures.

In this context the "Russian alloy" 40HNU with a composition of 57% Ni, 40%Cr, and 3% Al (in short: Ni57Cr40Al3 or simply NiCrAl) should be mentioned. Its chemical composition resembles some Inconel alloys such as the Inconel FM72, but it is not officially classified as superalloy since its high-temperature properties are not known. This material was extensively used in the former Soviet Union for high pressure equipment due to its exceptional strength. Eremets cites σ_Y=1.47 GPa and σ_T=1.57 GPa, A=15-20% for the material hardened to 45-48 HRC as well as σ_Y=1.96-2.06 GPa, σ_T=2.35 GPa, A=0-3% when hardened to 64-67 HRC [4]. It is still important for pressure cells which are used for magnetic measurements since Ni57Cr40Al3 is non-magnetic, i.e. paramagnetic with an almost temperature independent susceptibility of 4 x 10^{-8} m^3/kg at 300 K and 1 x 10^{-7} m^3/kg at 4 K [17]. Some of the cells discussed in chapter 5.2 and provided by Russian sources are made of this material. The alloy is unfortunately difficult to obtain, but a Japanese version might be available in the future [18]. In this material a small amount of boron of 50 ppm was added to improve the forging process. The hardening of NiCrAl (presumingly through precipitation) is carried out at 700-750°C which can therefore be taken as the upper temperature limit. If isotopically

TABLE 3.3

Typical chemical composition and mechanical properties of super-alloys Inconel 718 and PER706. Strength data correspond to bulk. To allow a meaningful comparison, data for Inconel 718 were taken from the equivalent grade PYRAD53NW from the same producer as PER706 (Auber & Duval).

Grade		718	PER706
Composition (mass%)	Ni	52	42
	Cr	19	16
	Nb	5	3
	Fe	20	37
	Mo	3	9
	Ti	1	1.8
	Co	<0.40	
	Al	<0.8	0.2
	Mn	<0.4	
Density (g/cm^3, 20°C)		8.22	8.1
Yield stress (GPa)	20°C	1.2	1.0
	600°C	1.0	0.8
Tensile stress (GPa)	20°C	1.4	1.3
	600°C	1.1	1.0
Elongation (%, 20-650°C)		20	19
Young modulus (GPa)	20°C	199	211
	650°C	166	175
Thermal conductivity (W/mK)		11	12
Lin. therm. exp. coeff. (10^{-5}/K, 300K)		1.3	1.4

enriched ^{11}B is used then NiCrAl could become an interesting material for neutron scattering applications on a wider scale, including to moderately high temperatures.

3.8   Hydrogen compatibility

Hydrogen is able to penetrate easily into steel and cause a massive decrease in toughness and strength, an effect known as hydrogen embrittlement. This can lead to catastrophic failure of high pressure equipment in contact with hydrogen. The issue of H-embrittlement is highly relevant for high pressure neutron cells.

   Generally speaking all steels carry a considerable risk of hydrogen embrittlement, but the magnitude of the effect depends strongly on the type of steel, its heat treatment (strength), the physical conditions of temperature, pressure, and time of H_2-exposure as well as the strain the metal is subjected

TABLE 3.4

Steels and non-ferrous alloys ranked by their
susceptibility to H-embrittlement, after [3], based
on data from NASA. See Table 3.2 for details.

Extremely attacked:	410, 440C, 430F
	17-4PH, 17-7PH
	Maraging steels
Severely attacked:	Nickel 200, 270
	Inconel 625, 70, 718
	Rene 41
	Hastelloy X
	Waspalloy, Udimet 700
	Ti-6Al-4V
	Ti-5Al-2.5Sn
Moderately attacked:	304, 305, 310
Not attacked:	316, 347, 660
	CuBe 25
	Al & Cu-alloys
	K-Monel

to. Contrary to widespread belief, the use of stainless steel is no guarantee
for avoiding hydrogen embrittlement. There is however a strong tendency for
austenitic steels to resist much better to H-embrittlement than martensitic
and ferritic steels, due to the high mobility of hydrogen in the bcc structure.
Spain & Paauwe [3] give a ranking of H-embrittlement risk for various types
of steels, based on data from NASA, and which is reproduced in Table 3.4.
Important observations are: (1) High-strength steels are extremely susceptible
to H-attack, including the stainless steels of the 400 and 630 AISI series (Table
3.2). Within a given class of steels the hydrogen embrittlement increases dramatically with strength, as clearly demonstrated on PH-type stainless steels
[19]. (2) The only proven safeguards against H-embrittlement are the stainless
austenitic steels of the 300 AISI series, in particular the 316 which is used for
hydrogen-proof gas handling systems, as well as the precipitation hardened
austenitic AISI 660 (A-286, nr. 1.4980). Latter is used as membrane material
in hydrogen-compatible compressors, as well as high-pressure hydrogen auto-
claves. None of these materials counts as high-strength steel, though the latter
can reach a tensile strength of 1.0 GPa at 300 K. (3) It is clear that CuBe is
unaffected by hydrogen, even at high pressure in the kbar-range.

A possible option against hydrogen attack is plating, i.e. the protection
by thin metallic films, a method which has been studied in depth. For this
purpose the steady flow J of hydrogen at pressure P across a material in form
of a plate is measured. It follows the relation [20]:

$$J \propto \sqrt{P} \exp(\alpha/RT) \qquad (3.1)$$

i.e. the penetration of hydrogen increases with the square-root of pressure and

shows an Arrhenius-type temperature dependence. A useful quantity is hence
the hydrogen permeability Π:

$$\Pi = \frac{JL}{A\sqrt{P}} \qquad (3.2)$$

where A is the surface and L the thickness of the plate. Its unit is hence cm^3
gas$\cdot cm^{-1}\cdot s^{-1}\cdot bar^{-1/2} = cm^2\cdot s^{-1} bar^{-1/2}$.

Figure 3.3 shows H permeabilities of various materials [21], including
steels. As expected from Table 3.4, the austenitic (γ) steels are much more
resistant to H-uptake than the ferritic and martensitic α-steels. For the pure
metals the permeability decreases in the order Ni-Pt-Cu-Mo-Al-Ag-Au-W.
Whereas the H-permeability of Ni is approximately one order of magnitude
lower than for an alloyed steel, it is found 10^3 lower for Pt and 10^9 lower for
W. Palladium is an exceptional case with its well-known affinity to hydrogen.
These data indicate that plating steels with some of these metals would be an
effective H-protection. Plating ferritic stainless steels with a 10 μm layer of
tin was found to reduce the permeability at 500 K by at least 10^4 [20]. Also,
oxide films of TiO_2 and Al_2O_3 with a thickness of 50-500 Å have been found
to be very efficient barriers to hydrogen, as concluded from a measured loss of
ductility by only 5-10% compared to 80% for the non-coated specimens [19].
It is interesting to note that an oxide film might grow on certain steels by its
own during the aging process in an oxygen-rich atmosphere. The unexpected
resistance to H-embrittlement of some high P/T cells might be an effect of
such self-grown oxide barriers.

To resume, it is clear that hydrogen penetration in steels can be drastically
reduced by various surface barriers. In practice, however, such surface treatment of high pressure equipment is prone to damage and then presents only a
limited protection to H-embrittlement. The safest way to avoid hydrogen embrittlement is hence still the choice of the appropriate alloy. Copper-beryllium
is probably the best choice, supported by the data in Figure 3.3 (pure copper)
and extensive practical experience with high pressure equipment in contact
with hydrogen.

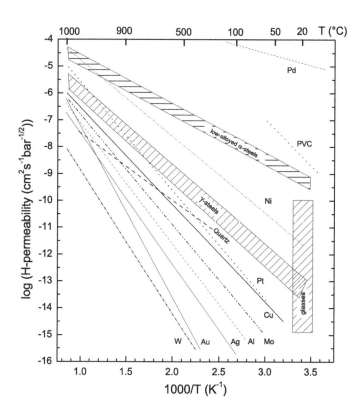

FIGURE 3.3
Hydrogen permeability (equation 3.2) for various materials. Adapted from ref.
[21].

4

![line decoration]

Construction materials III: Sinter materials

CONTENTS

4.1 Tungsten carbide (TC)

4.1.1 General

This class of materials should correctly be called *cemented carbide*, but is usually simply termed "tungsten carbide." We will use this expression throughout this book and abbreviate it as "TC" to distinguish it from pure tungsten carbide (WC) which is its principal constituent. It denotes a composite (hence an inhomogeneous) material made of tungsten carbide grains and a metal binder, in general Co, and in some cases Ni. Cemented tungsten carbide is used in high pressure devices for its exceptional compression strength, its relative incompressibility (high Young modulus), and its resistance to wear. Typical applications in high pressure devices are:

- Anvils and backing plates (multi-anvil and opposed anvil devices, belts, and Drickamer cells)

- Pistons in piston-cylinder cells, in particular clamp-type devices

- Liners in piston-cylinder cells, girdles, and belt-type cells

TABLE 4.1

Structural and physical properties of pure WC.

Space group	$P\bar{6}m2$ (No. 187)				
Lattice parameters (Å)	a=2.903	c=2.837			
Positions	atom	x	y	z	occ
	W	0	0	0	1
	C	2/3	1/3	1/2	1
Density	15.77 g/cm^3				
Melting point	2776 °C				
Thermal conductivity	29.3 W/mK (300 K)				

An important fact is that the mechanical and physical properties of TC can be varied over a wide range by controlling essentially two parameters: (a) the Co (or Ni) content, and (b) the grain size. This will be discussed in detail in the following paragraphs.

4.1.2   Production

The production of cemented tungsten carbide is a typical sintering process. The starting material is a powder of WC and Co (Ni) which was prepared by a special grinding process in a liquid environment which contains an organic solute to give the desired particle size. The powder is then compacted using a mold designed to the shape of the end product, and either hydraulic or rotative presses to generate the force. Sometimes isostatic compression to 0.2 GPa is applied to improve the homogeneity of the compacted powder. For the production of parts in small quantities, blocks or cylinders are produced from which the end product is machined by grinding or electro-erosion. The density of the material after compacting is approximately 7.5-8.5 g/cm^3, which is approximately 52-55% of the end product. The next step is the sintering process which involved heating to 450°C in an Ar of H atmosphere to eliminate the residual traces of organic binder. At temperatures close to 1300°C the WC starts to dissolve in the Co by solid state diffusion. Further heating leads eventually to the melting of Co (Ni) and elimination of the voids as well as to a considerable decrease of linear dimensions. The final density is 13-15 g/cm^3, depending on the Co content (Figure 4.1), with a porosity of usually less than 0.02%. The parts are then cooled back to ambient temperature over a period of 15-48 hours where the dissolved WC precipitates onto the existing WC particles, which leads to a cemented compound containing at 300 K almost pure WC grains embedded in a compact matrix of almost pure cobalt. Table 4.1 gives crystallographic details of WC.

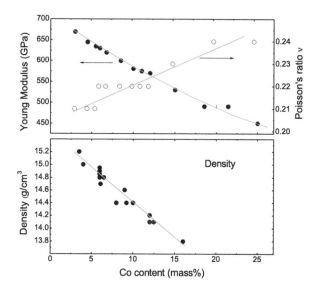

FIGURE 4.1

Density, Young modulus, and Poisson's ratio of TC as a function of Co content in various grades produced by Kennametal (some of them are listed in Table 4.2). Source: Kennametal France.

4.1.3 Mechanical and physical properties

The density of pure tungsten carbide is 15.77 g/cm³, that of cobalt 8.90 g/cm³. The density of TC decreases hence significantly as the Co content increases, see Figure 4.1. Tungsten carbide has a very high Young's modulus of 450-700 GPa, i.e. 2-3 times larger than steel (180-200 GPa), and it depends again strongly on the Co-content. In contrast, Poisson's ratio varies much less with the fraction of Co. One of the most important properties for high pressure applications is its compression strength which depends not only on the Co-content but to a large extent the grain size: the smaller the grain size, the higher the compression strength, see Figure 4.2. For standard TC grades with a Co content of 10% and average grain size, a rule of thumb is that the compression strength is approximately 50 tn/cm². This is 4-5 times larger than for high tensile steels. Not surprisingly the toughness shows the inverse behavior: low Co-containing carbides with small grain size have very low toughness, i.e. fracture easily. There is hence a trade-off between hardness and fracture toughness, and the choice of the grade depends strongly on the application. For items which suffer little flexion, because they are sufficiently supported for example, TC grades with high compression strength are clearly preferable. Examples are steel supported cylindrical backing plates. For anvils, such as used in opposed-anvil devices, the situation is different, since the material

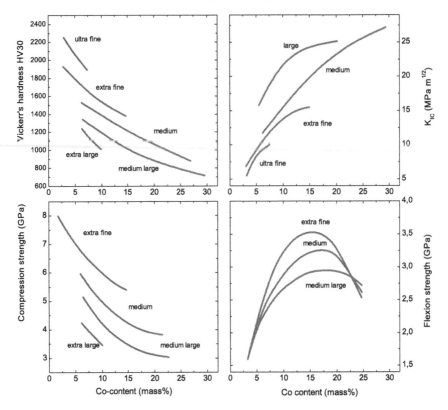

FIGURE 4.2
Typical dependence of mechanical properties of TC as a function of Co content
and grain size. Grain sizes vary from approximately 0.5 μm (ultrafine) to 5
μm (extra large). Source: Sandvik France.

is solicited under strong shear which can easily lead to fracture if the anvil
material is not sufficiently ductile.

Table 4.2 gives a list of TC grades available from Sandvik including some
of their mechanical and physical properties. The last letter in the name of
the grade designates the grain size. It ranges from typically 0.2 μm to >5 μm
and is divided in 7 classes: ultrafine (UF, <0.5μm), extra fine (F, 0.5-0.9μm),
fine (M, 1.0-1.3μm), medium (M, 1.4-2.0μm), medium coarse (P, 2.1-3.4μm),
coarse (C, 3.5-5.0μm), and extra coarse (T, >5μm). For high pressure applications it is mainly the UF, F, M, and P grades which are relevant. The
same Table 4.2 gives grades available from Kennametal (formerly Widia Hard
Materials), another major carbide producer. Note that there are grades where
Ni replaces Co as binder. These are sometimes called "antimagnetic" grades
since these materials are paramagnetic and the magnetic moment of Ni is

only 0.5 μ_B. Cemented tungsten carbides with Ni-binder do not activate under neutron irradiation, contrary to Co-containing material which produces the radioactive ^{60}Co isotope. Also, Ni-containing TC is less corrosive than the Co-containing grades. For all these reasons the Ni-containing grades have become the preferable choice for anvils used in high pressure neutron scattering applications, though Ni is a strong scatterer and can produce additional peaks in the diffraction patterns, see for example Figure 14.10.

4.1.4 Neutron attenuation

Figure 4.3 shows measurements of the attenuation of two different grades of TC as listed in Table 4.2. One is a standard grade containing 10% cobalt, the other is a grade with 9% nickel. It is the latter grade which is used for anvils in neutron scattering experiments with PE-type presses. The replacement of Ni by Co increases the absorption length by approximately 20% at 3 $\mathring{A}$ which translates to a gain of 26% in intensity across a 6 mm anvil. An inconvenience is that Ni is a strong neutron scatterer (in contrast to Co) which causes Bragg reflections in addition to those of WC.

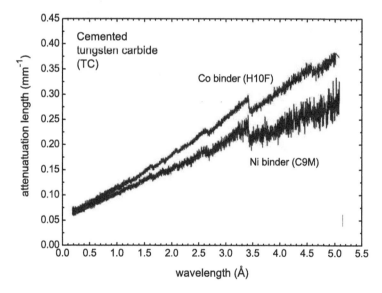

FIGURE 4.3
Attenuation of tungsten carbide containing Co and Ni as binder as a function of wavelength. Source: University of Edinburgh/Paris (unpublished).

TABLE 4.2

Selected TC grades available from Sandvik (upper) and Kennametal (lower) relevant for high pressure devices. R_{comp}: compression strength (in GPa), R_{tr}: transverse rupture strength (in GPa), HV30: Vickers hardness, T: toughness (in MPam$^{1/2}$).

Grade	WC (%)	Co (%)	Others (%)	ρ (g/cm³)	Grain size	E (GPa)	R_{comp}	R_{tr}	HV30	T
3UF	96.4	3.0	0.6	15.5	UF	670	11.5	2.1	>2250	5
6UF	93.1	6.0	0.9	14.75	UF	630	9.0	2.4	2050	8
8UF	91.1	8.0	1.0	14.5	UF	600	7.7	2.3	1900	10
H3F	97.0	3.0	-	15.3	EF	670	7.8	2.2	1925	7
H6F	93.7	6.0	0.3	14.9	EF	630	6.9	3.55	1775	10
H10F	89.5	10.0	0.5	14.5	EF	580	6.3	3.96	1600	13
H3M	96.5	3.5	-	15.2	F	670	6.9	1.7	1750	9
C9M	91.0	-	9 (Ni)	14.6	F	600	5.0	2.6	1375	10.7
C6N	94.0	-	6 (Ni)	14.95	M	620	4.8	1.75	1500	10

Grade	WC (%)	Co (%)	Others (%)	ρ (g/cm³)	Grain size	E (GPa)	R_{comp}	R_{tr}	HV30	T
KMS	89.3	10.0	0.7	14.4	0.8μm	600	5.5	3.6	1550	10.8
KTR	89.8	9.0	1.2	14.4	0.4μm	600	-	2.6	1840	9.0
THM-U	93.0	6.1	0.9	14.7	0.7μm	610	7.5	3.5	2020	8.3
K05C	93.4	6.0	0.6	14.8	0.6μm	610	6.0	2.8	1850	8.4
KF1	93.7	6.0	0.3	14.95	0.8μm	620	6.0	3.0	1740	9.0
K10	92.4	6.0	1.6	14.85	1.0μm	620	5.7	2.6	1690	9.3
K20	92.4	6.0	1.6	14.85	1.5μm	630	5.7	2.5	1640	9.3
KXC	93.5	6.0	0.5	14.90	2.6-5.0μm	630	5	3.1	1570	9.6
AFY	91.4	0	8.6 (Ni/Cr)	14.70	1.0μm	-	-	2.6	1550	-
GTD	90.3	0	9.7 (Ni/Cr)	14.50	2.5μm	570	5.5	2.6	1490	10

TABLE 4.3

Some physical properties of cubic boron nitride (cBN) sinter (Kiborit II)

cBN content	84%
Grain size:	10-40 μm
Density:	3.35-3.38 g/cm^3
Knoop hardness (at 10 N load):	28-30 GPa
Fracture toughness:	10.5 MPa m$^{1/2}$
Compression strength:	2.9 GPa
Thermal conductivity:	70 W/(mK)
Thermal stability:	up to 1400 K

4.2 Sintered cubic boron nitride (cBN)

This superhard material has recently gained considerable importance as anvil material in opposed anvil cells. Boron is a strong neutron absorber and anvils made of this material are more or less 'black,' depending on the wavelength of the neutron radiation used. In combination with null-scattering TiZr gaskets clean patterns can be obtained in angle-dispersive diffraction on continuous neutron sources. There are various superhard boron rich materials commercialized by essentially three companies, DeBeers, General Electric, and Sumitomo. The products available from these sources are known as Amborite and DBC (DeBeers), Borazon (General Electric), and BN (Sumitomo). Sintered items of cBN are either attached to a tungsten carbide substrate (DBC), or free standing in form of discs, and triangular and rectangular plates (Amborite). Starting with such products, the machining of anvils suited for neutron scattering applications is very costly. For this reason the material used so far for neutron applications is known under the trademark Kiborit. It can be obtained in final anvil shape for a cost which is not very much higher than that of tungsten carbide and considerably lower than that of sintered diamond. Kiborit is made of cBN powder and typically 10-15 mass% aluminum as starting material. This mixture is compressed to 4-5 GPa and simultaneously heated to 1200-1500°C where the Al melts and reacts with the cBN to give a sinter which consists of approximately 84% cBN plus AlN and AlB$_2$, as well as traces of pure Al. This material is non-conductive and cannot be machined by spark-erosion (EDM). Machining is hence only possibly by mechanical grinding or laser ablation. Table 4.3 gives some of its properties.

Although cBN is a strong neutron absorber, small diffraction lines can be visible for short wavelengths where the penetration depth becomes appreciable. Figure 4.4 shows three diffraction patterns obtained at a wavelength of 0.5 Å compared to simulations. For longer wavelengths the penetration depth decreases rapidly and Bragg reflections become almost invisible for wavelengths of standard diffraction experiments, i.e. 1-2 Å.

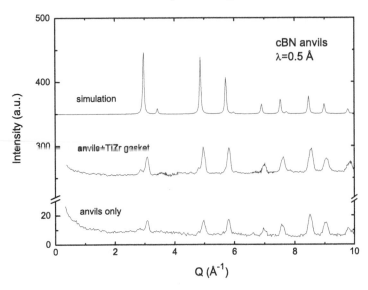

FIGURE 4.4
Diffraction patterns of cBN toroidal anvils, with and without a TiZr gasket in place in comparison with a simulated pattern (above). Data collected at the D4 diffractometer of the ILL. Source: Ph. Salmon, private communication.

4.3 Sintered diamond (SD)

4.3.1 General

This material is also called poly-crystalline diamond (PCD) and is used as anvil material for neutron scattering to very high pressures beyond 15 GPa. The main problem with SD is the commercial availability in sizes which are useful to neutron scattering, i.e. allowing large sample volumes (i.e. beyond 1 mm^3 and up to 100 mm^3). Contrary to tungsten carbide (TC), the number of producers is very limited and their focus is mainly on cutting tools which imposes geometries which are not adapted for high pressure applications. Apart from its superior strength SD shows larger transparency to thermal neutrons compared to TC.

Sintered diamond is made of 70-90% (in volume) diamond powder and an additive, in most cases cobalt, in some cases SiC or ceramics. The additive is a sintering aid, usually called "binder". The latter term is however misleading since it is not the binder which holds the diamond grains together, i.e. if the binder would be removed (which is possible by chemical edging), the remaining material would still have very similar mechanical properties. The compressive and transverse fracture strengths are in fact largely independent of the type of sintering aid but instead governed by the sintering process, the grain size, and the degree of densification [22]. Compared to tungsten carbide (TC), the particle size of the major component (diamond) is considerably larger, i.e. between 1 and 25 μm.

Sintered diamond is opaque in the visible and has a dark-grey appearance. Material with Co or SiC binder is electrically conducting which allows machining by spark erosion. The exact fabrication method is not revealed to the public, but what is known is that the sintering process involves very high pressures of several GPa and high temperatures, carried out in belt-type presses. As a consequence, sintered diamond parts cannot be produced in any size and shape, contrary to the case of tungsten carbide. The most convenient product for high pressure applications are cylindrical wire-drawing dies made of a core of sintered diamond and a mantle of TC as shown in Figure 4.3.3. This SD material contains Co as binder and will here be denoted as Co-SD, to distinguish it from the less easily available SiC-SD material, also commercialized under the name "advanced diamond composite" (ADC).

4.3.2 Co-Sintered diamond wired drawing dies

Such dies are available in diameters of 10-45 mm and thicknesses of 5-25 mm, approximately, and have therefore the size needed to make anvils and backing discs for large volume high pressure devices. There are currently only two sources of such dies, Diamond Innovation (formerly General Electric, now part of Sandvik) and Sumitomo Hard Materials. The corresponding products carry the name COMPAX and SYNDIE, and the cost of a COMPAX 5913 die with 32 mm overall diameter, 20 mm core diameter, and 13 mm thickness is approximately 1000 Euro. To this has to be added the machining costs to produce anvils, which can be considerable. Figure 4.3.3 shows examples of raw and machined COMPAX dies. The advantage of the TC mantle is that its machining is relatively simple and inexpensive compared to SD which means that the diameter of the die can be easily adapted to the specific application. The lateral faces can easily be shaped with a conical angle needed to provide a radial support (frettage).

4.3.3 SiC-Sintered diamond composites

This material has become available recently in the form of free standing dies of 20 mm diameter (Figure 4.3.3). The material uses SiC as binder which increases considerably its transparency to thermal neutrons (see below) and avoids activation through the production of ^{60}Co. The market for this material being very small, the production of SiC-SD has so far been pushed only by small companies on a more or less experimental scale. For this reason the experience with this material for high pressure neutron applications is rather limited. What we present here are data obtained on SiC-SD dies provided by Almax Industries in 2004 shown in Figure 4.3.3. This material has been carefully characterized in terms of its mechanical properties (Table 4.3.4) and neutron transparency (section 4.3.5). These data demonstrate the considerable potential of this material for high pressure neutron techniques and will hopefully lead to its continuous and sustained production and availability in the future.

FIGURE 4.5
Dies of sintered diamond and sintered cubic boron nitride (cBN). (1) Raw blank of COMPAX 5913 tungsten carbide supported sintered diamond (Diamond Innovations Inc.); (2) Double-toroidal anvil die made of tungsten carbide supported Co-sintered diamond blank type SUMIDIA WD960 (Sumitomo); (3) Raw blank of free standing SiC-sintered diamond (Almax Industries); (4) single-toroidal anvil die made of cBN Kiborit.

4.3.4   Mechanical and physical properties

Similar to the situation in TC, the physical and mechanical properties of SD are strongly dependent on the grain size of the diamond component. Figure 4.6 shows the diamond content as a function of grain size between 1 and 45 μm for Co-SD, a similar relationship holds for SiC-SD. As the grain size increases the diamond content increases from 83% to 95%. Since diamond has a density of 3.54 g/cm^3 and cobalt of 8.90 g/cm^3, the density of the sinter decreases with increasing particle size. Since the binder is the main source of neutron absorption, in particular in the case of Co, the neutron transparency increases with increasing particle size. Unfortunately the strength of SD is highest for small particle sizes, i.e. large binder content. For applications as anvils for neutron scattering the Co-SD wire drawing dies (core diameter 15-20 mm) are only available with grain sizes of 25 μm. Note that the transverse rupture strength of such material is by a factor 2 smaller than that of tungsten carbide, see Table 4.2, i.e. SD cannot support sever flexion, irrespective of what binder is used. We have used over the last 15 years products of both providers mentioned above and have not found any difference in mechanical performance. The numbers given in Table 4.4 are those available for COMPAX dies and might not differ very much from those of SYNDIE. Note that the compressive strength of this material is in the order of 8 GPa and hence close to the value stated by Horton for SiC-SD [22] which confirms that the type of binder has a minor influence on the mechanical properties of the composite. The numbers for SiC-SD in Table 4.3.4 correspond to a product commercialized by Almax Industries and differ not too much from the data of Horton et al. [22], given the considerable variation of the mechanical properties with the sintering process and the difficulty to measure them accurately. They should be regarded as representative for the whole class of SiC-SD composites.

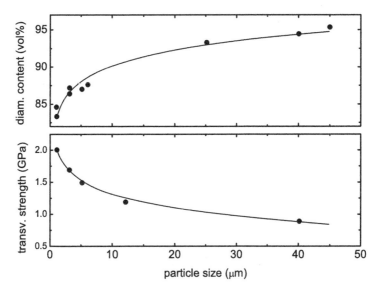

FIGURE 4.6
Diamond content and transverse compression strength of polycrystalline sintered diamond (Co-SD). Source: Sumitomo.

COMPAX and SYNDIE dies have the tendency to develop micro-cracks under load, for example when used as anvil material. The same is most likely the case for other types of SD as this problem has already been reported for freestanding SiC-SD blanks used in multianvil devices [23]. These cracks are in general invisible and do not lead to a visible fracture of the whole anvil, but compressed samples with low shear strength such as liquids and soft metals can easily escape through them. For anvils made of wire-drawing dies such

TABLE 4.4

Some mechanical and physical properties of
COMPAX 5913 PCD dies from Diamond
Innovations. Source: Diamond Innovations.

Diamond content (vol)	94% (balance Co)
Grain size	25 μm
Density	4.18 g/cm^3
Young modulus	913 GPa
Compressive strength	7.6-8.3 GPa
Trans. rupture strength	1.5 GPa
Knoop hardness	$\approx$36 GPa
Thermal expansion	4.2 10^{-6} K^{-1} (SYNDIE)
Thermal conductivity	5-6 Wcm^{-1}K^{-1}

TABLE 4.5

Mechanical and physical properties of SiC-SD
from Almax.

Diamond content (vol)	70-75% (balance SiC)
Grain size	25/6 μm (bimodal)
Density	3.4 g/cm^3
Young modulus	860 GPa
Poisson's ratio	0.09
Compressive strength	4.7 GPa
Trans. rupture strength	0.5 GPa
Knoop hardness	$\approx$62 GPa
Fracture toughness	9.7 MPa m$^{1/2}$
Thermal expansion	2.7 10^{-6} K^{-1}
Thermal conductivity	300 Wm^{-1}K^{-1}
Electrical resistivity	0.05 Ω/cm

microcracks can even travel across the SD-TC boundary and reappear at the
surface several mm within the TC mantle.

4.3.5   Neutron related properties

Besides strength a considerable advantage of sintered diamond is its high
neutron transparency, see Figure 4.7. At a wavelength of 3 Å, Co-SD has an
attenuation length $\approx$ 2 times larger than cemented tungsten carbide (TC)
(Figure 4.3), and for cobalt free SiC-SD material it is even 4 times larger
at this wavelength. To give an example, for a neutron beam passing through
10 mm COMPAX anvil material, the increased transparency translates in an
expected gain in signal of approximately 4.5 compared to standard TC anvils.
As a consequence, a reduction of sample volume with the aim to reach higher
pressures is not accompanied by a reduction in data quality. Note that the
smaller absorption and the fact that diamond is a strong neutron scatterer

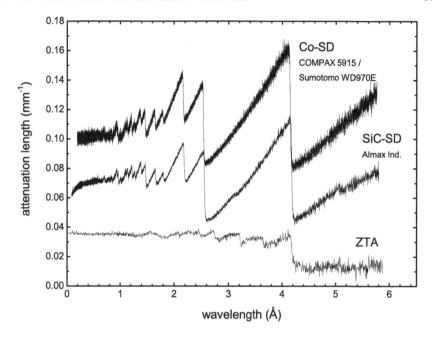

FIGURE 4.7
Neutron attenuation of sintered diamond containing Co and SiC binders (Co-SD and SiC-SD), as well as zirconia-toughened alumina (ZTA). Bragg edges at 4.1 Å, 2.52 Å and 2.06 Å correspond to the first three Bragg reflections (111), (220) and (222) of diamond (a=3.57 Å). The weak Bragg edges at 5.03 Å and 3.05 Å in the SiC-SD material are due to the (111) and (220) reflection of SiC (a=4.36 Å). Data for ZTA from W.G. Marshall, ISIS.

leads to strong Bragg edges (see Figure 4.3 for a comparison with TC) and hence a considerable distortion of a polychromatic neutron beam which has to be corrected. Given the fact that the higher costs (factor $\approx$4) is partially outweighed by the increased life time of SD anvils, it is foreseeable that the use of sintered diamond in high pressure neutron scattering will become more and more popular, even for a pressure range where TC could be used.

When used in angular dispersive diffraction SD produces strong and sharp Bragg reflections which might interfere with the sample signal. However, due to the small lattice parameter and the high crystal symmetry, the number of observable reflections using standard wavelengths between 1 and 3 Å is very limited. As shown in Figure 4.8, for a wavelength of 1.87 Å there are only three diamond reflections in the pattern, and none beyond 4.06 Å there is none. The strongest reflection of the binder material, Co, is hardly detectable. The situation is slightly worse with SiC-SD which produces several additional weak reflections, in the case of λ=1.87 Å at least three. It should be noted though that an oscillating radial collimator (ROC, such as available at D20, ILL and HRPT, PSI) will reduce the diamond and SiC reflections drastically since they occur generally (depending on λ) at 2θ values between $40°$ and $140°$ where the ROC is most efficient. This is clearly seen in the pattern of cBN anvils, where the two reflections visible at $Q=4.9~\AA^{-1}$ and $Q=5.75~\AA^{-1}$ are barely observable in measurements where the ROC is in place.

4.4   Perspective: Nano-polycrystalline diamond (NPD)

This is a new diamond material produced by direct conversion of graphite under pressure and temperature conditions of 12-25 GPa and 2300-2500 K [24]. It is a dense bulk material containing granular diamond crystals of 10-20 nm. The material is transparent and has a Knoop hardness of 110-140 GPa and is therefore as hard as single crystal diamond, if not harder. One of the key advantages is that it is isotropic and hence not susceptible to cleavage like single crystals. Anvils made of NPD have been used in diamond anvil cells to reached pressures beyond 100 GPa [25]. Recently dies of 10 mm diameter were produced by the Matsuyama group, and it is believed that dies of up to 15 mm can be made with the currently available high pressure equipment. It is foreseeable that anvils made of this material will be used for neutron scattering and could probably allow compression of samples in the mm^3 size to 50 GPa or beyond.

4.5   Ceramics

Certain ceramics are useful in high pressure neutron cells due to their high neutron transparency combined with high strength and low thermal conductivity. For example high-density Al_2O_3 has been used as cylinder material in the McWhan cell for the last 4 decades. Recent advances in ceramic technology have resulted in materials with strength close to that of TC. In Table 4.6, mechanical properties of two high-strength materials are listed which find applications in high pressure devices. Yttria-stabilized zirconia has a compression strength of approximately half of TC, but a thermal conductivity one order of magnitude lower. It is routinely used as backing seats in high P/T neutron cells where it serves as thermal insulation of the anvils. A more recent material is zirconia-toughened alumina (ZTA) which has a compression strength of 4.7 GPa, and a fracture toughness of 6-7 $MPam^{1/2}$, hence mechanical properties close to those of TC (Table 4.2). The outstanding advantage of alumina-based ceramics is their neutron transparency, in particular at large wavelengths.

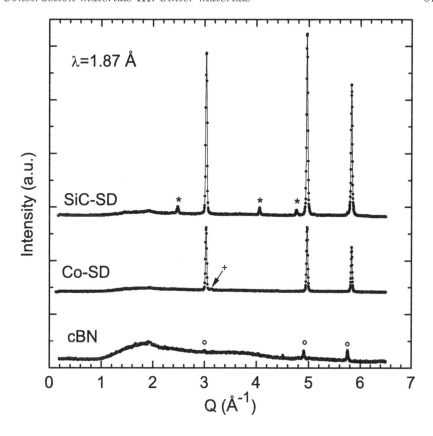

FIGURE 4.8
Diffraction patterns of toroidal anvils (no gasket). Lower: cBN anvils, (○) indicated Bragg reflections from cBN. Middle: Co-SD anvils (COMPAX 5913), strong reflections are from diamond, (+) is the only visible reflection of Co. Upper: SiC-SD (Almax); strong reflections are from diamond, (∗) from SiC (space group F$\bar{4}$3m, a=4.36 Å). All data recorded at the D20 diffractometer (ILL) without oscillating radial collimator and normalized to same monitor count. Anvils were all single-toroidal with an identical profile and a gap of 1.5 mm between them, mounted in a VX5 Paris-Edinburgh pressure cell at 300 K, in air. Co-SD and SiC-SD pattern were shifted vertically by an equal amount to avoid overlap. Source: S.K., unpublished.

Figure 4.7 gives a comparison of attenuation lengths of ZTA with diamond, see Figure 4.3 for a comparison with TC. The exceptional strength of this material allows toroidal anvils for Paris-Edinburgh cells to be made, which have been successfully used since mid-2011 at ISIS in the 0-7 GPa range

TABLE 4.6

Typical properties of high-strength ceramics ZrO_2-Y_2O_3 and Al_2O_3-ZrO_2 (zirconia-toughened alumina, ZTA). Source: CeramTec.

Grade	ZrO_2-Y_2O_3	Al_2O_3-ZrO_2
Composition (mass% ZrO_2)	≈ 95	25
Density (g/cm^3)	6.08	4.37
Compression strength (GPa)	2.2	4.7
Flexure strength (GPa)	1.4	1.35
Toughness K_{IC} (MPam$^{1/2}$)	6.7	6.4
Hardness, HV1	1350	1760
Young modulus (GPa)	210	357
Poisson's ratio	0.30	0.24
Thermal conductivity (W/mK)	2.5	17
Lin. therm. exp. coeff. (10^{-6}/K, 300K)	11.2	8.1
Specific heat (J/gK)	0.4	0.7

(W.G. Marshall, private communication). For a wavelength of 4 Å the increased transparency translates to a gain in diffracted signal by one order of magnitude.

5

Liquid/gas and clamp pressure cells

CONTENTS

5.1 Liquid/gas pressure cells

5.1.1 General

The characteristic of these types of cells is that they use a liquid or gaseous pressure transmitting medium which is injected by a compressor through a high pressure hose, usually a capillary. The cell is hence permanently (i.e. at least for the length of the measurement) connected to the compressor. The pressure can be changed in situ, and measured easily by standard manometers attached to the compressor-side of the equipment. This is hence the ideal tool when fine-tuning of pressure is required. The pressure conditions are per se hydrostatic, guaranteed by the choice of the adequate pressure transmitting medium, since otherwise the equipment does not work properly (blocked capillary). The fluid of choice is obviously helium since its freezing pressure (Appendix E) is higher than any other gas/fluid. For low-temperature measurements (below typically 100 K) this choice becomes unavoidable. However helium is able to penetrate into many compounds, the best known case is ice to form clathrates. The compatibility of the pressurizing fluids used in liquid and gas cells is hence an important consideration which has to be clarified well before the start of an experiment. Helium has also the well known inconvenience of being difficult to seal, in other words its viscosity is so low that even small leaks can become a serious problem, in particular at low temperatures. The use of more standard fluids would hence be desirable but is in practice rarely possible since (a) it has to be compatible with neutron measurements (no hydrogen, low neutrons absorption), and (b), it has to have a high freezing pressure, even at low temperatures. In practice, only other rare gases can be considered, such as argon, neon, and sometimes nitrogen.

5.1.2 Examples

In this chapter a few examples of gas/liquid cells are given. These specific designs were selected since (a) the device are frequently used in standard experiments on user-facilities, its reliability and life-time is hence well established. (b) the devices exhibit typical design elements which might be useful for readers who want to build similar cells which are more adapted to particular applications.

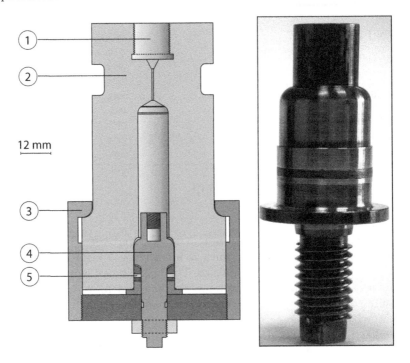

FIGURE 5.1
Gas pressure cell for pressures up to 0.7 GPa used at the LLB. (1) Gas inlet, (2) body, (3) retaining cap, (4) mushroom plug, enlarged views shown in the picture to the right, (5) Bridgman seal. Courtesy F. Maignen, LLB, Saclay.

Figure 5.1 shows the 0.7 GPa gas cell with inner diameter of 12 mm and a maximal outer diameter of 57 mm at the bottom. Since the pressure chamber has a height of 40 mm, the maximum available sample volume is 4.5 cm^3 which is a convenient quantity for most inelastic neutron scattering studies. The relevant cylindrical part has an outer diameter of 48 mm, the ratio K is therefore 4. The cell is made of high tensile Aluminium alloy 7049-T6 (chapter 2.4) having a yield stress of 0.5 GPa. The cell is autofretted with the elastic-plastic interface located at a diameter of approximately 18 mm. Taking these data, the safe working pressure is approximately 0.54 GPa, according to equation 1.30 and the rupture pressure is 0.80 GPa according to equation 1.28. In a test carried out on this cell at 300 K, rupture occurred at 8364 bar. The top of the cell contains a fitting where the capillary is connected. To avoid leaks, the capillary is rarely disconnected. The sample is loaded from the bottom, and the cell is sealed by a mushroom Bridgman-type seal. Gas pressure is generated by a standard gas-intensifier stage using a membrane compressor as first stage. This cell is frequently used for inelastic measurements down to 1.8 K, mostly using helium, in rare cases nitrogen.

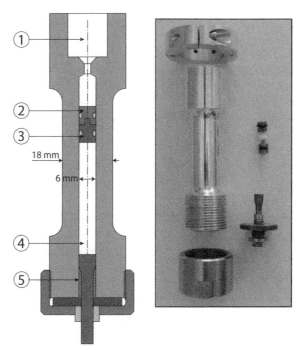

FIGURE 5.2
Liquid pressure cell made of 7049A-T6 aluminum alloy for soft condensed matter research at ILL. (1) fluid inlet, (2)-(3) floating barrier, (4) sample space, (5) mushroom plug with Bridgman seal. Source: ILL, Grenoble.

Figure 5.2 shows a liquid cell made of 7049A-T6 alloy and mainly used for soft condensed matter research at the ILL [26]. It is based on an earlier design made of CuBe and which is used at the LLB to pressures up to 0.7 GPa [27]. The available sample volume is approximately 1 cm^3. As in the LLB aluminum cell (see above), the high pressure capillary inlet is attached at the top end and the sample is introduced from the bottom, and closed by a mushroom-type Bridgman seal. This cell is autofretted to 0.5 GPa, and the value K=18/6=3.0 would indeed predict a rupture pressure of 0.7 GPa at 300 K if a yield strength of 0.55 GPa is assumed, a value which seems slightly optimistic, see chapter 2.4. Note however that the strength of high tensile Al-alloys is strongly temperature dependent (chapter 2.4). Whereas it increases by at least 20% from room temperature to 4 K, it drastically *decreases* beyond 100°C, i.e. these types of cells can only be used for moderate high temperatures. The particularity of this cell is the use of a floating barrier which separates the compression liquid (Fluorinert, above the piston) from the sample. A detailed description of such a barrier was given by Susman and Kleb [28] which applied it in a 0.7 GPa 7075-T6 cell for diffraction studies on liquids and glasses.

The TiZr gas cell shown in Figure 5.3 is conceptually very similar, i.e. having at one side a standard 3/16"-tube cone-cone fitting with a M16 gland nut for the gas inlet and a Bridgman mushroom plug at the other. The plug is pre-compressed by an external nut (item 6) which secures initial leak-tightness for gases even at low pressures and low temperatures. Given a ratio of outer to inner diameter of K=28/7=4, the rupture pressure is approximately 1.0 GPa according to equation 1.28 and taking a yield strength of 0.65 GPa for TiZr (2.4).

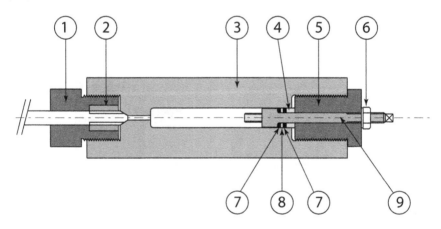

FIGURE 5.3
TiZr gas pressure cell in use at ISIS for pressures up to 0.6 GPa. (1)/(2) high pressure gas inlet (3/16" tube), (3) body, (4) backing ring, (5) seal plug, (6) mushroom nut, (7)/(8) seal (Cu/lead/Cu), (9) mushroom plug. Source: R. Done, ISIS.

Figure 5.4 shows a high temperature gas cell made of Inconel used at ISIS. Given the strength of the Inconel and a wall ratio of K=21/7=3 the bursting pressure of this cell is approximately 0.8 GPa at ambient temperature and 0.5 GPa a 500°C. The cell fits into the standard ISIS furnaces, and it is only the front (part) 120 mm, which is heated to the maximum temperature whereas the top including the gaskets and flanges remain essentially at room temperatures. The purpose of the stainless steel baffle flanges inside the cell is to reduce the gas convection as well as the dead volume. This cell has been used for experiments with standard inert gases but also with hydrogen under high pressure and temperatures. The hydrogen gas experiment was carried out to 0.2 GPa and 600°C without any detectable deterioration of the cell which suggests a certain degree of resistance of Inconel 718 to hydrogen embrittlement. An obvious difficulty in this cell is the strong attenuation and scattering of the cell material. The latter can nevertheless be suppressed in a 2-θ = 90° geometry where tight shielding of the incoming and diffracted beam is most efficient. The cell was used at the POLARIS diffractometer where such a scattering geometry is possible.

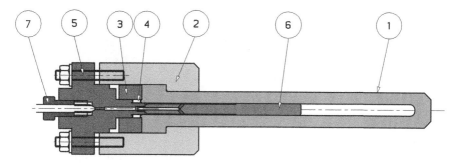

FIGURE 5.4
Gas pressure cell for high temperatures up to 700 K in use at ISIS. (1) body, (2) support cap, (3) backing ring, (4) gasket seal, (5) pressure cap, (6) baffle pin, (7) gas inlet (3/16" tube). Items (1)-(3) and (5) are made of Inconel 718, item (4) of pure Cu, item (6) of stainless steel (316). Source: R. Done, ISIS.

An example for a high pressure gas/liquid cell for single crystal crystallographic studies is given in Figure 5.5. This cell is made of TiZr and used at the ISIS SXD diffractometer for gas pressure up to 0.2 GPa. The outer and inner diameters are only 3.2 mm and 2 mm respectively, which indicates a bursting pressure of 0.38 GPa taking a yield strength of 0.7 GPa. The pressure transmitting medium is helium injected through a 1/16" gas tube with standard high pressure fittings. This allows hydrostatic pressure conditions down to the lowest temperatures attainable by the cryostat. The upper temperature limit is 300K or slightly above.

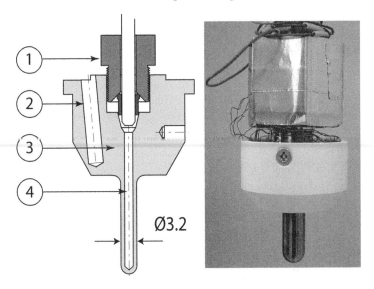

FIGURE 5.5
Gas pressure cell for single crystal diffraction at SXD (ISIS) up to 0.3 GPa.
(1) Gas inlet with 1/16" high pressure fitting, (2) hole for cartridge heater,
(3) body (TiZr), (4) pressure chamber. Source: R. Done, ISIS.

5.2   Clamp cells

5.2.1   General

A clamp cell is a piston-cylinder pressure device where the force necessary
to generate the pressure is applied by an external press and then locked me-
chanically. The pressure can hence not be changed in situ (on the beamline),
but only by removing the cell to increase the force on the piston. Clamps are
hence relatively compact and conceptually simple devices. They are per se
liquid cells, since it is difficult to load a clamp with a gas. Although this could
be done using a high pressure gas loader, in practice it would be complicated
due to the complex assembly which contains the sample. In general the sample
is not directly loaded into the cylinder but loaded first into a capsule together
with the pressure transmitting fluid. This assembly is then introduced into the
bore of the cylinder and compressed by two pistons on each side. The reason
for using a capsule is leakage: a liquid introduced directly without capsule
would unavoidably leak through the gap between piston and cylinder. The
material and sealing technique of the capsule varies from cell to cell.

   A general problem of clamps is the partial loss of force on the piston when
the load is decreased after pressure changes and tightening the locking mech-

anism. This loss of load can be limited by the use of a mounting rig which holds the clamp on its upper part during pressure increase. Pressures are determined at 300 K from the applied force and some correction factor which takes account of the friction and the loss upon unloading. But accurate pressure values have always to be determined by an in-situ measurement, for example by introducing a pressure marker with known equation of state (NaCl, Pb). Cooling a clamp-type cell also leads to a pressure loss due to the thermal contraction of the pressure medium which the cell body cannot fully compensate. Despite all these inconveniences compared to gas/liquid cells described in the previous chapter, clamp-type cells are the most convenient technique for pressures beyond typically 1 GPa. This is because compressors and intensifiers as used in gas/liquid cells become increasingly complex, expensive, and unreliable beyond this pressure.

5.2.2 Examples

In the following we give a few examples which illustrate the general principle and pressure performances of clamp cells.

Figure 5.6 shows a 1.2 GPa clamp cell made of fully hardened CuBe which is frequently used at the ILL. The inner and outer diameters are 5 and 19 mm, hence K=3.8, and the cylinder is autofretted. The upper part of the clamp has an increased diameter for a thread which allows the cell to be held on the upper part during the pressure increase. Without this mounting (i.e. by simply putting it on a flat surface) during pressure increase, the clamp body would elongate when the load is decreased after tightening the locking mechanism, which would lead to a considerable loss in pressure. The sample capsule used in this cell is made of aluminum with an Al cap with the sealing being achieved by two Cu-In plugs on each side. Two tungsten carbide pistons with conical tips push on each side. A similar cell exists also in null-scattering TiZr for pressures up to 1.2 GPa.

Figure 5.7 shows two clamp cells for pressures up to 1.5 GPa with inner diameters of 6-8 mm, hence for relatively large sample volumes in the cm^3 range. These cells are made of a compound cylinder, the two parts being made of high tensile Russian aluminum alloys. The inner part is slightly conical (typically 1°) and pushed into the outer cylinder to produce a radial support (frettage). There is also a steel liner. The load onto the TC pistons (not shown) is applied via a shaft introduced at the top of the cell and locked by tightening one of the two end nuts. The capsules used in this kind of cell are made of lead or aluminium, including the plug. The sample (in most cases single crystals) are inserted into the lead capsule which is filled with the pressure transmitting medium, in most cases Fluorinert, immediately closed by the plug, and compressed inside the cell to seal it. Single crystal inelastic scattering is the principal application of this cell.

Figure 5.8 shows a clamp cell design by Kamarád et al. which belongs to the standard high pressure equipment of the HMI neutron source (Berlin).

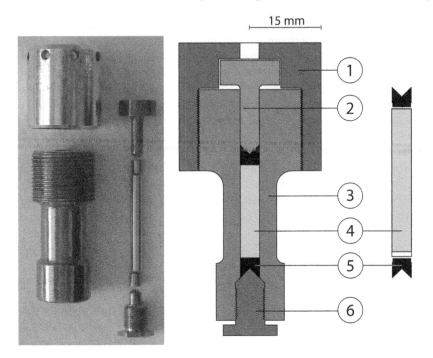

FIGURE 5.6
CuBe clamp cell for pressures up to 1.2 GPa used at the ILL. (1) retaining
cap, (2) piston (TC), (3) body (CuBe), (4) sample container (Al), (5) plugs
(In-coated CuBe), (6) fitting. The sample capsule on the right is shown in its
initial length. Source: ILL, Grenoble.

The maximal pressure is 1.0 GPa, the inner and outer diameters are 3.5 and 12
mm, hence K=3.42. The particularity of this cell is that pressure can be easily
measured in situ using a Manganin coil which is integrated in one of the plugs.
The cell is mostly used for measurements on magnetic systems and hence
compatible with existing high-field and low temperature equipment. In most
of such experiments the pressure determination could be done by diffraction
methods, using the equation of state of the sample itself (if known) or by
adding an adequate pressure marker. But for certain applications (experiments
on disordered systems, for example) the use of in situ pressure sensors might
be unavoidable and the design in Figure 5.8 offers one way to achieve this. Note
that this cell uses no sample capsule, i.e. the sample and pressure transmitting
medium are loaded directly into the bore of the cell. As mentioned in the
introduction, one of the disadvantages of clamp devices is the partial loss in
pressure under cooling. The "HMI-3.5" cell was used to evaluate this effect
using different pressure transmitting media, see Figure 5.9. Although only the

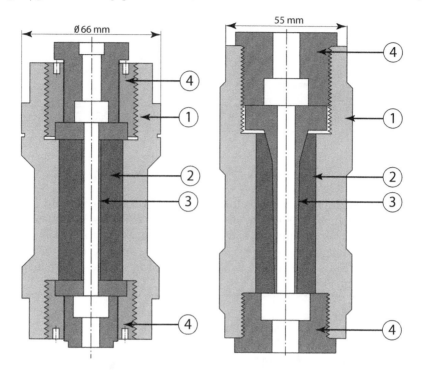

FIGURE 5.7

Clamp cells for inelastic neutron scattering. Left: HPC15 for pressures to 1.5 GPa. Right: PSI cell for pressure to 1.2 GPa. (1) body; (2) jacket; (3) liner; (4) locking nut. Compression pistons and sample capsules are not shown. Parts (1) and (2) are from Al-alloy (B95T & B96T respectively), part (3) and (4) of steel (45XMHΦA). Source: R. Sadikov, priv. communication.

FC77 Fluorinert liquid is useful for neutron scattering, the data give an idea about the size of the pressure loss.

Figure 5.10 shows a clamp cell for single-crystal diffraction measurements. This design can be regarded as the clamp-version of the single-crystal gas-cell shown in Figure 5.5, allowing larger sample volumes and higher pressures. The body of the cell is made of either null-scattering TiZr or of the high-tensile non-magnetic alloy "HNU" (Ni57Cr40Al3) (see chapter 3.7). The inner and outer diameters are 4 and 15 mm, respectively, i.e., K=3.75. The sample is located inside a capsule together with the pressure transmitting medium which is then inserted into the cell. Unlike most standard clamps, the body of this device has only one opening which simplifies the design but makes the extraction of the sample capsule more difficult. Pressure is applied by mounting the clamp on a rig and applying load onto the piston (item 2) using a shaft introduced through the bore of the locking nut (item 1). The load is then locked by tightening the

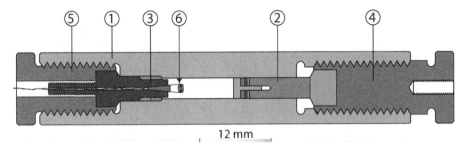

FIGURE 5.8
"HMI-3.5" high pressure cell for pressures to 1.2 GPa: (1) cell body, (2) piston
with Bridgman mushroom plug at front end. Seals: Cu-In-Pb-Cu packing, (3)
plug; seal: rubber O-ring seal and tinned Cu ring, (4) piston holder, (5) plug
holder, (6) Manganin pressure sensor (coil). All parts are made of hardened
CuBe, if not stated otherwise. Source: J. Kamarád, private communication.

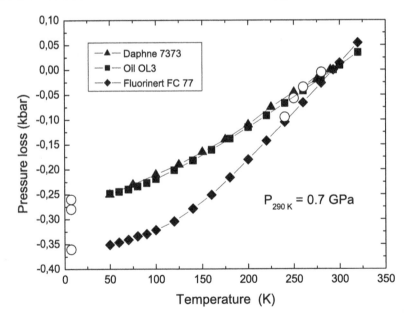

FIGURE 5.9
Pressure loss upon cooling using the HMI-3.5 cell (Figure 5.8) and different
pressure transmitting media. Filled symbols correspond to pressure measurements with the Manganin gauge, open symbols by measurements of T_c of Pb
(low T) and T_N of Lu_2Fe_{17}. Source: J. Kamarád, private communication.

nut. Using the HNU alloy pressures of 1.9 GPa have been attained, for TiZr the pressure limit appears to be approximately 0.5 GPa. Friction in this kind of design can be considerable and the applied force might suggest a higher pressure than in fact measured by in-situ methods.

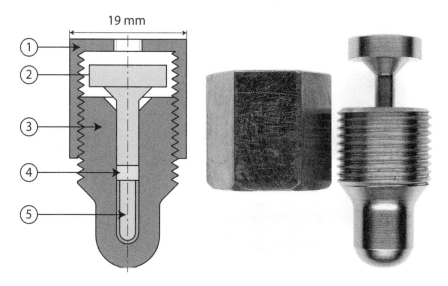

FIGURE 5.10

Miniature clamp for single crystal neutron scattering to 2 GPa: (1) locking nut; (2) piston; (3) body; (4) end plug; (5) teflon capsule. Parts (1)-(5) are made of high tensile steel (HNU). A TiZr version (item 3) is available for pressures to 0.5 GPa. Source: R. Sadikov, private communication.

6

McWhan-type cells

CONTENTS

6.1 General

McWhan and collaborators at Grenoble presented in 1974 a piston-cylinder device which became for the following 25 years the working horse for high pressure neutron scattering beyond ~ 1 GPa [29]. The characteristic element of the McWhan cell (Figures 6.1 and 6.2) is a biconical cylinder made of high density sintered Al_2O_3 alumina[1] ("Lucalox" from GE or "Widalox R" from Krupp) or ZrO_2 zirconia with a bore of ca 5 mm which contains the sample (plus other elements as discussed further below) and two pistons made of tungsten carbide on each side. Lucalox is composed of 99.9% pure Al_2O_3 and the rest MgO as binder and has a compression strength of approximately 3.8 GPa [32]. The cylinder is on both ends radially supported by a steel ring which is pushed over the two conical parts. In McWhan's original design [29], the device had two independently driven hydraulic rams: One to drive the upper tungsten carbide piston to compress the sample and another to push the support rings around the bicone. The force on the support rings was carefully adjusted as the pressure in the sample was increased. Using this method, the maximal reported pressure is 4.2 GPa, derived from the observed first-order phase-transition in Cs. The today's accepted value for this transition is still the same, so it is clear that the McWhan design is capable to generate pressures beyond 4 GPa.

The presence of two hydraulic rams made the original McWhan cell complicated to handle and incompatible with low-temperature experiments, which was one of the main scientific interests of the inventors. Later designs moved

---

[1] The idea of using pre-stressed Al_2O_3 cylinders for high pressure neutron scattering was not entirely new, see the work in the 1960s by Brugger, Worlton, and Bennion [30, 31] on Fe_2O_3 and Bi where diffraction data up to 4.1 GPa were collected.

therefore quickly towards clamped McWhan cells [32, 33, 34], which can be considerably miniaturized. Each change in pressure needs hence the device to be removed from the beamline and the use of an external press to increase the load. A design which used a carbon-fiber rod to increase the pressure in situ became never popular, possibly because of the insufficient amount of thrust which can be transmitted through a rod without flexion [35].

There are numerous versions of clamped McWhan cells of various sizes and for a range of applications. Two design concepts, the ISIS and ILL McWhan cells will be discussed in more detail below. In practice, clamped McWhan cells operate routinely up to $\sim$ 2.5 GPa, and rarely up to 3 GPa. The absence of adjustable radial support on the bicone of such clamped cells is undoubtedly the reason for its reduced pressure performance compared to the original design.

It is a well known observation that even at moderate pressures the bicone often fails by developing vertical cracks, but nevertheless holds the pressure and might even be reused [32]. Such a geometry of a cracked bicone is similar to that of the segmented cylinder, see section 1.2.4, for which the elastic limit is given by equation 1.33. Taking typical values of K=8 (bicone-compression pads) and σ_Y=1.4 GPa, equation 1.33 would predict P=2.5 GPa, which is indeed the typical pressure limit of standard McWhan cells. This suggests that considerably higher pressures could be achieved by increasing the outer diameter of the compression pads. McWhan's original design [29] had indeed K=16, which would indicate maximal pressures of $\approx$ 5 GPa. The limiting factor in this case was however not the bicone but the tungsten carbide pistons, a phenomenon well know to occur in piston-cylinder devices operating in this pressure range [5].

6.1.1 The ISIS McWhan-type cell

The choice of aluminium or zirconium oxide as pressure vessel is not based on its mechanical properties. On the contrary, the tensile strength of sintered Al_2O_3 is rather low, approximately 0.25 GPa [37], which explains the need for radial support along the biconical lateral faces. The use of this material derives from its very high neutron transmission. The absorption coefficient at λ=1.08 Å of Widalox R is μ=0.011 mm^{-1} [38], which gives a transmission of 80% across the entire bicone. A cylinder of similar dimensions made of high tensile steel would transmit only approximately 40%. Sintered Al_2O_3 is obviously completely non-magnetic, which makes the material useful for magnetic high pressure studies.

The considerable disadvantage of the McWhan cell for powder diffraction is the large number of Bragg reflections caused by Al_2O_3 which hampers seriously structural studies using angle-dispersive diffraction. An instructive example is given by Kuhs et al. who used a McWhan cell to investigate the structures of high pressure ice phases VI, VII, and VIII [39]. This problem can be avoided in time-of-flight measurements which allow a fixed diffraction

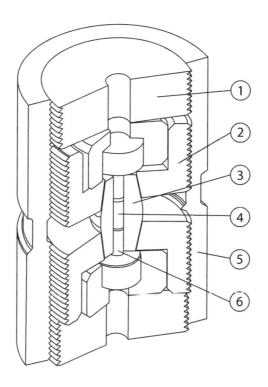

FIGURE 6.1

McWhan cell for time-of-flight measurements at ISIS. (1) locking pad, (2) compression pad, (3) sintered Al_2O_3 bicone, (4) pressure capsule, (5) Al body (7075-T6 alloy), (6) tungsten carbide piston, (7) pressure pad. Items (1), (2), and (7) are made of Vascomax 300 maraging steel. Redrawn from Ref. [36].

angle $2\theta=90$ deg, where the collimation of the incident and diffracted beam can be very efficient. To give an example, data on CuO [40] and CuI [41] demonstrate that it is possible to obtain clean diffraction patterns without any contaminant Bragg reflection from the bicone. Also, in single-crystal studies the background poses in general no particular problem as shown in ref. [42] in work on K_2HPO_4 to 1.7 GPa.

Figure 6.1 shows the design which is currently used at the ISIS facility for both diffraction and inelastic scattering [36]. It is conceptually the simplest clamp-type McWhan cell and was also designed by McWhan himself [33]. The initial preparation of the cell consists of pre-stressing the bicone. This is done by tightening the compression pads (2) onto the bicone. The threads on the pads have opposite sense, so by holding the two pads and turning the cell body, the pads will come closer to each other and compress the bicone. This is done under a press which applies an increasing force onto the pads as the cell body is successively turned using a handle. A lubricant is usually applied on the conical faces to reduce friction. This operation is stopped at a load of approximately 30 tonnes, the load is decreased and the sample introduced into the bore. This is done using a capsule containing a pressure transmitting medium, usually made of a thin-wall aluminum cylinder with an end cap and a CuBe-ring to seal it, very similar to the design used in clamps like the one shown in Figure 5.6. Traditionally Fluorinert type fluids are used since the hydrostatic limit at 300 K matches the maximum pressure which can be generated in such cells, i.e. approximately 2-2.5 GPa. Methanol-Ethanol mixtures have been used as well, but the general experience is that it is more difficult to seal and its compressibility is larger which reduces the available sample space. Also, the thermal contraction upon cooling is larger which leads to a larger pressure loss at low temperatures compared to other pressure transmitting fluids.

6.1.2   The ILL McWhan-type cell

The ILL-design is more complex since the clamp mechanism to pre-stress the bicone is separated from the clamp which pushes onto the WC-pistons to lock the pressure in the sample chamber. The two clamp bodies fit into each other like Russian puppets. This is illustrated in Figure 6.2. Items (2), (5), and (11) belong to the ("primary") clamp which compress the compression pads (9), locked by item (2) and (5). The "secondary" clamp which locks the force on the two tungsten carbide pistons (4, 8) consists of items (1), (3), and (7).

There are two versions of this cell, one with a 5 mm bore in the bicone to reach 3.0 GPa and a second model with a 7 mm bicone for pressures up to 2.0 GPa.

6.1.3   Other McWhan-type designs

As stated further above, there are numerous design concepts of McWhan cells, adapted to various types of measurements and sample environments, and a few examples are given here.

Bloch proposed a compact cell with 68 mm outer diameter and a mass of 4.5 kg which is compatible with standard low temperature equipment [32]. The maximal pressure using a NaCl sample was 3.4 GPa, the pressure loss between 300 K and 4 K typically 10%.

Onodera at al. [43] reported a design with 100 mm external diameter, 5 kg mass, and 8 mm bore diameter. This cell is similar to the ISIS-type cell but includes a different thrust mechanism for the compression pads. The pressure performance of this cell has been carefully studied by both diffraction

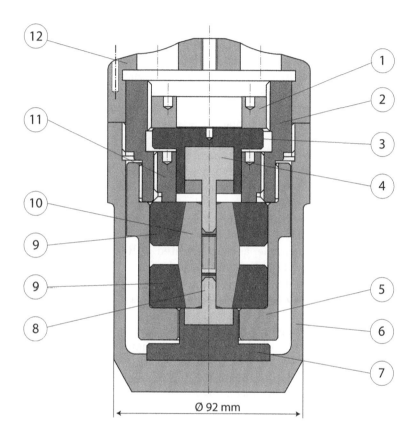

FIGURE 6.2
Cross section of the ILL McWhan-type cell. (1) locking ring, (2)/(5) inner clamp body, (3)/(7) support spacers, (4) backing discs, (6)/(12) outer clamp body, (8) pistons, (9) compression pads, (10) Al_2O_3 bicone, (11) locking ring (for compression pads). Items (1), (2), (3), (7), and (9) are made of MARVAL 18 maraging steel, items (5) and (6) of 7049 aluminum alloy, items (4) and (8) of tungsten carbide, and item (12) of 2017 aluminium alloy. Source: ILL, Grenoble.

methods and a Manganin coil inside the pressure chamber. These show that due to friction the pressure inside the sample chamber is typically 70% of the nominal pressure determined from the applied force and the dimensions of the tungsten carbide piston. The pressure loss upon cooling was also determined and is less than 10% between ambient temperature and 10 K. The cell was used with both Al_2O_3 and ZrO_2 bicones. A useful ancillary equipment is a miniature hydraulic ram which allows pressure changes without removing the cell from the beamline.

An interesting excursion from designs using oxides as bicone material is Knorr's "Kiel-Berlin" cell [44]. This McWhan-type cell was primarily built for diffraction on continuous sources. To avoid the aforementioned contamination from the pressure container, the bicone is made of null-scattering TiZr alloy, which is radially supported in the same way as all other McWhan-type cells. The "Kiel-Berlin" cell allows high-temperature measurements via internal heating using a graphite furnace. Pressures of up to 1.5 GPa at 700 K are reported.

A heatable McWhan cell for time-of-flight (TOF) measurements was presented by Hull et al. [45]. Similar to the "Kiel-Berlin" cell, internal heating is achieved by a graphite sleeve, but the bicone is made of Al_2O_3. In TOF measurements effective shielding is possible since the diffraction angle is fixed. Clean diffraction patterns can therefore be obtained despite the various elements (Al_2O_3 graphite, thermal insulation, electrical contacts) in immediate contact with the sample, as demonstrated in Ref. [45]. This pressure cell achieved 1.2 GPa and 1100 K simultaneously.

Walter reported a design with 60 mm diameter which is adapted to time-of-flight inelastic scattering up to 3.5 GPa [38]. This cell was optimized by finite element methods. An important contribution of this work is that it points out how the maximal achievable pressure depends on the size H of the unsupported part of the bicone, see inset Figure 6.3 for definitions. The relevant parameter seems to be the 'aspect ratio' $H/(D-d)$. The motivation for plotting the aspect ratio against P_{max}/K in Figure 6.3 is the evidence that the compression pads/bicone assembly operates at high pressure as a segmented cylinder (see remarks further above) for which $P_{max} \propto K$ with $K = \Delta/d$ (the global wall ratio of the assembly)². It is seen that data from various groups fall approximately on a straight line which for $H \to 0$ approaches indeed the value expected for the segmented cylinder if one takes σ_Y=1.4 GPa for the yield strength of the compression pads. The important message is that for given bicone and compression pad diameters (d, D, and Δ), the pressure performance decreases rapidly with increasing H and reaches a factor 0.5 when H is roughly three times the wall thickness $(D-d)/2$. And also, that the pressure performance of most McWhan cells could be improved if the exterior diameter of the compression pads Δ is increased while keeping H as small as possible.

---

²Note that Walter [38] plots the aspect ratio against P_{max}/lnK, with $K = D/d$ the wall ratio of the *bicone*. This seems unjustified since the lnK dependence applies only for plastic deformations (see sect. 1.2.3), which do not occur in a ceramic bicone.

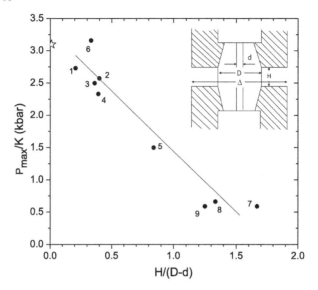

FIGURE 6.3

Empirical relation between the maximal achievable pressure P_{max}, global wall ratio $K = \Delta/d$ and the 'aspect ratio' $H/(D-d)$. Source: (1) McWhan et al. [29]; (2) Bloch et al. [32]; (3-4) ILL cells, priv. comm. J.-L. Laborier; (5) Hull et al. [45] (H=21 mm); (6-9) Bloch & Paureau [35]. The star indicates the limit for a segmented cylinder, equation 1.33. The line is a guide to the eye.

6.2 Conclusion

McWhan-type cells are nowadays less and less used, mainly because of their limited pressure range. However, in the 1-3 GPa range they might still be an interesting alternative to Paris-Edinburgh-type (PE) devices. They provide sample volumes of typically 500 mm^3 which is 5-10 times larger than the visible sample volume in toroidal anvils used in the PE press. Also, the piston-cylinder geometry of the McWhan cell allows almost any compression factor $\triangle V/V$ at constant signal level. This is not the case for cupped anvils used in PE-applications, where $\triangle V/V$ is limited to approximately 0.4 and the signal decreases by a factor ~ 3 between ambient and maximum pressure, see chapter 10. A known inconvenience of the McWhan cell is that the bicone is usually lost after each experiment, i.e. for each experiment a new bicone has to be fitted. The reason is that either it breaks already during the pressure increase, close to the maximum attainable pressure, or it breaks when the locking pads have to be taken off to release the pistons and recover the sample. As a result, the loading procedure tends to be time-consuming and repetitive experiments might be expensive, since the machining of the various capsule elements and possibly also the pistons have to be included.

7

Sapphire, moissanite, and diamond anvil cells

CONTENTS

7.1   Sapphire cells

What we call here "sapphire cells" are clamped panoramic cells which have a capacity of 5-25 tonnes and which are usually equipped with large gem-stone anvils of 5-15 mm diameter. In practice, anvils of this size are almost always sapphire, but may as well be of moissanite or zirconia. There seems to be no clear agreement on the naming of such cells. What is clear is that historically the first design emanated from the Kurchatov Institute in Moscow where such cells were made of CuBe and used by Somenkov and collaborators for high pressure neutron scattering at the local reactor. The technology was then exported to the Laboratory Léon Brillouin LLB (Saclay, France) by Somenkov and Goncharenko who used them extensively at the Saclay neutron source ORPHÉE for low-temperature powder neutron scattering measurements and called them "Kurchatov-LLB cells" from the mid-1990's on. Apparently parallel to these developments Ahsbahs and Kuhs in Germany experimented with sapphire anvils of various shapes with the aim to carry out single-crystal neutron diffraction measurements on ferroelectrics, both at ambient and low temperatures. These cells were also clamps with a panoramic view onto the sample to maximize the available Q-space, and hence resembled in many aspects the Russian design, in particular the later design which uses sapphire spheres as anvils. The Russian neutron scattering community at the Dubna reactor continues to use these cells and calls them simply "sapphire cells," and we adopt this nomenclature throughout this book.

7.1.1   Thrust mechanism

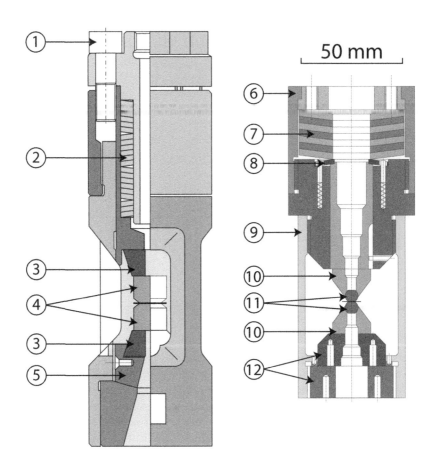

FIGURE 7.1

Cross section of clamped sapphire cells. Left: The 4-column Kurchatov-Dubna cell equipped with cylindrical anvils. Right: Ahsbahs' cell for single crystal diffraction using spherical anvils. (1) compression bolt, (2) Belleville washers, (3) anvil seats, (4) anvils, (5) rocking/alignment mechanism; cell made entirely of CuBe alloy. (6) compression cap (steel), (7) Belleville washers, (8) anti-twist ring, (9) cylinder (Al-7075 tube), (10) anvil seats (Thyradur-steel), (11) anvils (sapphire spheres) (12) backing plates (steel). Adapted after D. Kozlenko and H. Ahsbahs, private communication.

Figure 7.1 gives a cross section of the Dubna and Ahsbahs designs, and the picture in Figure 7.2 compares the former with Goncharenko's "Kurchatov-LLB" version. All three are conceptually very similar: The diameter is large enough to have a thrust of at least 5 tonnes required to deform gaskets over a surface of typically 1-5 mm diameter, but small enough to be used with standard ("orange") cryostats or high magnetic fields. All of them have large azimuthal apertures to allow single crystal investigations, and all of them use bevel washers to maintain the force on the anvils when the cell is cooled. The force is generated by manually tightening 6 bolts which compress the washers. The Kurchatov versions have a rocking mechanism which allows the lower anvil to be aligned, as well as set-screws for horizontal translation of the anvils. Aligning the sapphire spheres in Ahsbahs' cell is trivial as explained further below and there is no need for an aligning mechanism. The cell currently used at Dubna has 4 tie rods which gives a horizontal aperture of 4x70°. This cell is adapted to time-of-flight measurements at the DN-12 diffractometer [46] at a diffraction angle of 90°, i.e. the incident beam travels along the thrust axis, crosses the upper anvil, and the diffracted neutrons are collected by detectors in the equatorial plane of the anvils. The "Kurchatov-LLB" cell is adapted to a continuous source and angle-dispersive diffraction which needs large angular apertures in the equatorial plane which explains the two-column design.

Apart from the standard cell shown in Figure 7.2, Goncharenko presented two smaller as well as one larger version, all made of CuBe [47]. The smallest one ("egg cell") had a capacity of $\approx$ 3 tonnes (30 kN) and its overall dimensions did not exceed 16 mm which allows the cell to be turned horizontally in a cryostat. The other two cells are scaled-down and scaled-up versions of the standard design, with external diameters of 32 mm and 67 mm respectively, and capacities of 5 and 25 tonnes. The smaller of these versions allowed use in dilution cryostats to temperatures down to 0.1 K. Both this and the egg cell could fit into the LLB 7.5 T magnet. The impact of these three other cells appears to be limited. The reason is probably that the small cells offer only small sample volumes or low pressures due to their limited capacity, and the large cell could not be used neither in standard cryostats nor magnets due to its considerable diameter. With these constrains the real advantage compared to Paris-Edinburgh type cells is no longer obvious.

7.1.2 Sapphire anvils

Sapphire anvils have been used since decades in opposed-anvil devices because of some advantages in its optical properties compared to diamond. Sapphire is transparent in the wavelength range 144 nm - 6 μm, i.e. from the ultraviolet to the infrared, i.e. does not show the intrinsic absorption edge at 220 nm of diamond nor absorption of nitrogen impurities which prevent measurements below 330 nm for the majority of diamond anvils (type Ia). Sapphire is also preferable when the Raman peaks under investigation fall in the range of the one- and two-phonon Raman signal of diamond. Its application for neutron scattering is mainly due to the fact that large gems of typically 10 mm diameter are available at almost negligible price which allow therefore sample volumes in the mm^3 range, orders of magnitude larger than for diamond anvils

FIGURE 7.2
Goncharenko's two-column Kurchatov-LLB cell (left) and the four column Kurchatov-Dubna design (right). Both are made entirely of CuBe-alloy. Photograph courtesy of N. Rey and D. Kozlenko.

of standard size. The pressure performance of large sapphire anvils as a function of different anvil geometries, sample volumes, orientation of the gems, gasket materials, and pressure transmitting media has been investigated in detail and allow clear guidelines to be given for applications for high pressure neutron scattering.

Generally speaking sapphire is despite of its considerable hardness a fragile anvil material and surface flaws such as scratches can easily occur and lead to premature failure of the anvil under pressure. Without surprise this effect is amplified when hydrostatic pressure media (liquids) are compressed, and it will be seen that the pressure performance in such cases is systematically lower than for the case of solid pressure transmitting media. For this reason the initial material for making the anvils must be of high quality. A priori Czochralski-grown material has less defects but most of the commercially available sapphire is nowadays made by the Verneuil process. Eremets [4] recommends initial annealing of the material at 1900-1950°C for 6-8 hours to remove residual stresses, as well as low-temperature annealing at 1200°C during 1 hour after polishing to eliminate surface cracks. This procedure apparently increases the strength of the material by a factor 2-3, but there seem to be no systematic high pressure studies to support this. Most of the material commercially available, in particular the sapphire spheres, appears to be already of high bulk quality, as judged from inspection by optical techniques.

A major advantage of transparent anvils is that pressure can be conveniently measured by the ruby fluorescence technique. However, in the case of sapphire anvils some care has to be taken since most sapphires (certainly the sapphire spheres) contain small amounts of Cr^{3+} impurities which cause a fluorescence which interferes with that of the ruby marker. The problem can be avoided by careful focalization of the laser beam, or by the use of a different pressure marker, such as SrB_4O_7:Sm^{2+} or $SrFCl$:Sm^{2+}.

Figure 7.3 shows three popular anvil geometries (as already shown in Figure 7.1) in some detail. There is the geometry (a) where the anvils are cut from a cylindrical rod of diameter D and provided with a cone of angle α with respect to the equatorial plane, and a culet of diameter d which is either flat or might have a spherical cup. Goncharenko [47] used dimensions of typically D=8 and 12 mm, the Dubna group somewhat larger gems of D=18 mm, but both with α=30°. Furuno reports dimensions of D=5 mm and α=18° to reach pressures of 14 GPa for optical applications [48]. Eremets reports that an additional bevel angle of 2° between the cone and the culet, similar to the bevels in diamond anvil cells, increases the pressure performance and pressures of 12 GPa were regularly achieved even using helium and argon as pressure transmitting medium [4]. Glazkov and Goncharenko [49] also used beveled anvils. In whatever case, there is strong evidence that removing the edge between the flat culet and cone/bevel is important for the strength of sapphire anvils. Patselov et al. [50] suggested that for a maximum pressure of 15 GPa the radius of the "rounded bevel" should be approximately half the radius of the tip.

As for the size of the culet, Goncharenko recommends that its area should not be larger than approximately 15 times the area of contact with the seat [47]. Also, in this design the hole through the seat is relatively small, i.e. 1-2 mm diameter for anvils with D between 8 and 12 mm, and 2-3 mm for anvils with 18 mm diameter.

As will be discussed further below, the gasket thickness for sapphire anvils is hardly larger than 0.2 mm if pressures of 5-10 GPa are aimed at. To increase the sample volume "cupped" anvils have been used, i.e. anvils with a spherical indentation, see Figure 7.3. The diameter is approximately the same as the gasket hole and the depth approximately 1/5 of the diameter. With these dimensions the sample volume can be easily doubled, probably

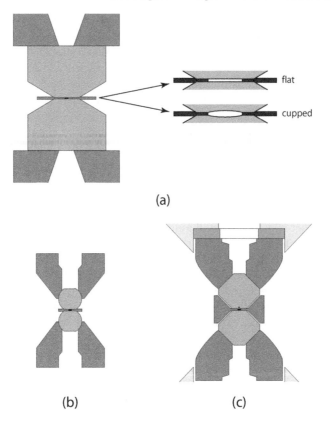

(a)

(b)   (c)

FIGURE 7.3
Anvils-gasket assembly in sapphire cells. (a) conical anvils with flat and
"cupped" culets, details are shown at the right; (b) spherical [51] and (c)
double-conical anvils [52], both with flat tips.

at the expense of pressure performance. The additional effect of the cupped
anvils is that this geometry reduces non-hydrostatic pressure conditions since
the compression is less uniaxial compared to flat anvils. Nevertheless, measurements of the pressure distribution inside such cupped anvils by the ruby
fluorescence technique and using no pressure transmitting medium indicate a
pressure variation of up to ±20% across the sample [53]. This value depends
however strongly on the sample, as explained in chapter 13.4.5. Experience
with cupped anvils (irrespective of the material) shows that they break easily in a purely hydrostatic environment. Given the requirement to maximize
the sample volume, almost all of the current and past neutron applications
of sapphire anvils, including the work by Goncharenko et al., were done in a

non- or quasi-hydrostatic environment, i.e. using a solid pressure transmitting medium (such as NaCl) or none at all.

In most of the cited work anvils with the crystallographic c-axis parallel to the thrust direction were used. But contrary to widespread belief and intuition, the crystallographic orientation of the sapphires with respect to thrust axis has no influence on the pressure performance. This is clearly demonstrated by the work of Furuno et al. [48] and consistent with our own experience on spherical sapphire anvils (see below). However, an orientation with the thrust axis (hence direction of view) along c or parallel to the hexagonal plane has the advantage of minimizing the effect of birefringence, which leads to images of bad quality apart from other disadvantages like the optical depolarization.

Another geometry is based on sapphire spheres which are commercially available in large dimensions, up to typically 12.5 mm (1/2 inch). The main advantage is cost: perfectly polished spheres of 1/4 inch diameter and high quality can be obtained for less than 50 Euros, and their performance appears to be not much different to the conical geometry discussed above. The spheres need to be provided with a flat culet if gaskets with holes larger than 0.1 mm are used, and obviously to give optical/visual access to the sample chamber. Takano et al. used culet diameters of 1.5-2.5 mm for spheres of 1/4 inch and reached under non-hydrostatic conditions almost 11 GPa [54]. Beveling of the culet seems not to have been applied yet, but "rounding" the edges is a standard procedure and highly recommended. For neutron applications only the single-crystal application of Ahsbahs is known in which pressures of 1.7 GPa were reached at relatively large volumes of 1 mm^3. As mentioned above, the orientation of the thrust axis seems to have no or only a minor influence on the strength of the anvil.

Apart from the low cost of sapphire spheres a considerable advantage is that the seating of such anvils is almost trivial, i.e. there is no real need for a rocking mechanism. For this purpose one of the anvils is glued into the spherical backing seat with an approximate alignment given by the cell. The second anvil is then inserted and the two culet brought into contact with each other. This will rotate the second anvil and self-align the two culet faces. The two anvils are left in this position till the glue has hardened.

Ahsbahs experimented in single-crystal neutron diffraction experiments with "double-conical" anvils, i.e. anvils which are seated on a conical surface. This particular seating prevents failure on the back of the anvil. Since the machining of such anvils is expensive, such an anvil geometry has never become popular. Note that these anvils were combined with gaskets which support the front faces of the anvils over a large area. Although such a gasket geometry increases the required load to obtain a certain pressure, it insures a well defined gasket geometry which is advantageous for attenuation corrections in single-crystal diffraction measurements. Flat gaskets such as shown in the Figure 7.3 tend to flip up- or downward and thus give an ill-defined gasket geometry which is difficult to model.

7.1.3   Gaskets

There is the general and experimentally well supported consensus that sapphire anvils can only be used with relatively soft gasket material, i.e. with metals of a yield strength of well below 1 GPa. Possible choices are: copper beryllium (ref. [55] and S.K., unpublished), pure copper [48], bronze (a copper-tin alloy) [54], Cu-Ni alloys [47], brass [56], aluminum (or rather a ductile Al-alloy) [47], and TiZr [51]. The last two materials are particularly important to neutron applications. Copper beryllium is usually used without heat-treatment, for example the "1/2 H" state. In experiments in our group using sample volumes of 1 mm^3 we use 1/2 H CuBe annealed at 188°C for 30 minutes which increases the yield strength to approximately 0.8 GPa (Figure 2.1). CuBe gaskets which are hardened beyond this value systematically lead to anvil failure. This observation supports Eremets' statement that stainless steel seems to be the hardest gasket material usable for sapphire anvils [4]. For neutron applications null-scattering TiZr is in principle the material of choice. Its tensile strength is 0.8 GPa and hence potentially compatible with sapphire anvils. However, TiZr has a strong tendency to stick to the anvil surface which leads frequently to premature anvil failure. Covering the anvil with an Al-surface (sputtering) definitely removes this inconvenience. The requirement of soft gasket materials means that the gaskets have to be thin which limits the available sample volume. Takano used a composite gasket with a 100 μm Inconel sheet sandwiched between two bronze sheets of 50 μm each [54]. The final thickness at 11 GPa was 100 μm compared to 70 μm for a simple bronze gasket of 100 μm at the start. A particular case is aluminum, or rather aluminium alloys such as the 7075 alloy (Table 2.5). The yield strength of sheets (hence cold worked material) can attain 0.7 GPa which is approximately at the limit of what can be sustained by sapphire. In addition, Al has a very low absorption and incoherent cross section. A gasket of 6 mm diameter absorbs only approximately 6% of a neutron beam which crosses its full diameter, i.e. the losses through attenuation are negligible. Since the absorption of the anvils are negligible as well, the visible sample volume is almost identical to the volume without pressure cell, even if the cell is tilted. Aluminum alloys have hence been the preferred gasket materials in the neutron work at LLB and Dubna. Goncharenko mentions though that liquids cannot be contained in Al-gaskets, but CuBe or Cu-Ni alloys can [47]. For time-of-light measurements with the beam along the thrust axis, a composite gasket is used by the Dubna group. There is an inner ring ("liner") of null scattering Ni-V with a wall thickness of approximately 1 mm which is surrounded by a gasket of Al-alloy. This setup insures diffraction patterns which are free of contaminant Bragg peaks from the gasket material.

TABLE 7.1

Typical anvil-gasket geometries from five selected sources. con./flat: conical with flat culet; con./cup with "cupped" culet. t_i: initial gasket thickness (in mm); $\varnothing_h/\varnothing_c$: diameter gasket hole/culet (in mm); Si-oil: silicon oil; graph: graphite; Cu/Inc/Cu: Cu-Inconel-Cu sandwich. PTM: pressure transmitting medium. P_{max} is in GPa.

Ref.	anvils	gasket	t_i	$\varnothing_h/\varnothing_c$	PMT	P_{max}
(T)[a]	spher.	bronze	0.10	0.35/1.5	Si-oil	12.6
	spher.	bronze	0.30	0.25/2.5	Si-oil	10.7
	spher.	bronze	0.10	1.0/2.5	KI	5.4
	spher.	Cu/Inc/Cu	0.20	0.25/1.5	Si-oil	11.8
(F)[b]	con./flat	Cu	0.30	0.4/1.0	graph.	0.02
(E)[c]	con./flat	brass	0.15	0.2/0.6	He	12
(K)[d]	spher.	CuBe	0.60	1.3/3.5	Si-oil	2.0
(G-K)[e]	con./flat	Al	0.50	0.8/1	NaCl	7.7
	con./cup.	Al	0.50	1.5/2.5	NaCl	7.5

[a]Takano et al. (Ref. [54]); [b]Furuno et al. (Ref. [48]); [c]Eremets [4]; [d]Klotz (unpublished); [e]Goncharenko (Ref. [47]), & Kozlenko (priv. comm.).

7.1.4 Pressure performance

The pressure performance as a function of volume and other parameters has been carefully studied, not only motivated by neutron applications. Figure 7.4 resumes the results obtained by five independent groups, for both conical and spherical anvils: (1) Gocharenko's data obtained on cupped conical anvils of various size under nonhydrostatic conditions (NaCl pressure transmitting medium in probably most cases); (2) Furuno et al's. data point using conical anvils with flat tips and non-hydrostatic conditions (graphite PTM) [48]; (3) Takano et al's data using spherical sapphire anvils with flat tips and, except one point, silicone oil and the 4:1 methanol-ethanol mixture [54]; only the points with the highest pressure performance were included; (4) Eremets' data point at 12 GPa using conical anvils and Ar and He as PTM; (5) Unpublished data by Klotz et al. using spherical sapphire anvils to 2 GPa and liquid water (to 100°C) as well as Daphne silicon oil as samples. Goncharenko's result indicate a maximum pressure which varies with the sample volume V approximately as $\propto 1/V^6$, and the same appears to be true for the other data sets. However, the pressure performance of Goncharenko seems to be considerably higher. There are two reasons for that: first, for a given pressure the sample volume can be larger due to the cupped culet geometry; and second, the experiments are all non-hydrostatic.

Figure 7.4 gives also an interesting comparison with the performance of single crystalline diamond anvils having the usual brilliant cut. Although the maximal attainable pressure is considerably larger, the decrease in pressure performance with volume is much more rapid. It is believed that this is due to flaws which are quite frequent, if not unavoidable, in large diamonds needed

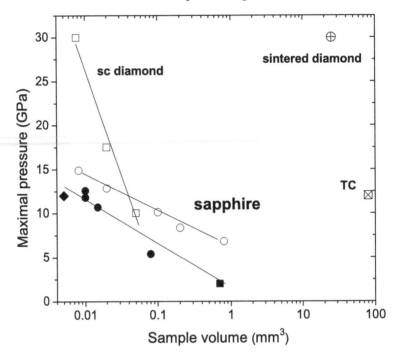

FIGURE 7.4
Pressure performance of sapphire anvils as a function of available sample volume, adapted from Ref. [47] ○: Goncharenko [47] and Kozlenko [priv. comm.] (conical anvils, non-hydrostatic), ●: Takano et al. (spherical anvils, hydrostatic) [54], ◆: Eremets [4] (conical, hydrostatic), ■: Klotz et al. (unpublished, spherical, hydrostatic). For comparison data for single-crystalline diamond anvils (□), sintered diamond (⊕), and tungsten carbide (TC, ⊠) are also given.

to compress relatively large samples. As a result of that, sapphire anvils with sample volumes larger than ~ 0.1 mm^3 perform better than diamond anvils. Obviously, for larger sample volumes, sapphire cannot compete with opaque sintered materials such as TC, cBN, or sintered diamond which allow easily sample volumes of 80 mm^3 to be compressed to 10 GPa.

From all these considerations it becomes clear that for neutron experiments the field of applications of sapphire anvils is very narrow: There has to be a good reason for optical access (pressure determination is not a good reason since it is easily measured by various other pressure markers like NaCl or Pb) and the physical problem to be studied has to be below 7 GPa. For higher pressures single crystalline diamond is favorable.

TABLE 7.2

Some mechanical and physical properties of gemstones used for anvils in high pressure cells.

	Sapphire	Moissanite	Diamond
Composition	Al_2O_3	SiC	C
Space group	R$\bar{3}$c (167)	P6$_3$mc (SiC-6H)	Fd$\bar{3}$m
Density (g/cm³)	3.97	3.217	3.52
Young modulus (GPa)	370-494	447[e]	1140[d]
Poisson's ratio[d]	0.27-0.30	0.17[c]	0.07[d]
Hardness (Knoop, GPa)	15-21	19[f]	80-140[a]
Hardness (Mohs)	9.0	9.3	10
Tensile strength (GPa, 300 K)	0.4	0.565	0.560
Fracture toughness K_{IC} (MPa m$^{1/2}$)	2.4-4.5	1.8[f]	5-7[b]
Index of refraction	1.768/1.760	2.65	2.42
Birefringence (n_o-n_e)	0.008	0.05	0
Specific heat (300 K, J/gK)	0.84	0.86	0.96
Thermal conductivity (W/cmK)	0.30-0.33	0.153	20
Lin. therm. exp. coeff. (10^{-5}/K, 300K)	0.5-0.66	2.3	0.11

[a]Ref. [57], depending on direction; [b]Ref. [58]; [c]Ref. [59]; [d]Ref. [60], aggregate average; [e]Ref. [61], aggregate average; [f]Ref. [62]; data for sapphire are from Ref. [63] and Ref. [64] (fracture toughness); ranges indicated anisotropy; index of refraction is at 589 nm.

7.2   Moissanite anvil cells (MACs)

The perspective of using moissanite (SiC-6H) as anvil material for neutron
diffraction was triggered by Xu and Mao's report to have reached 51 GPa
with gems of size and geometry similar to those used in diamond anvil cells
[65]. Since moissanite can be synthesized and is commercially available in large
dimensions equivalent to 300 carat diamond (typically 10 mm diameter), its
use for high pressure neutron scattering appears to be very promising. At first
glance the vastly superior pressure performance compared to sapphire seems
to be surprising when comparing the hardness and fracture toughness of the
two materials, which are not tremendously different, see Table 7.2. Nevertheless, there is sufficient experimental evidence that moissanite can indeed
support substantially higher pressures than sapphire, in particular under non-
hydrostatic conditions. The strong difference in pressure performance between
hydrostatic and non-hydrostatic conditions is already indicated by Xu and
Mao's data: whereas the reported maximum pressure of 51 GPa corresponds
to direct compression of a mixture of ruby and diamond powder, compression
with a methanol-ethanol fluid allowed only a maximum pressure of 10 GPa
[65].

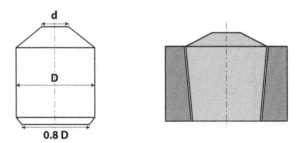

FIGURE 7.5
Freestanding (left) and radially supported (right) moissanite anvils, after Xu
et al. [66]. Culet diameters d are 0.2-4 mm, anvil diameters D are 3-10 mm.

The use of large volume moissanite anvils suited for neutron scattering is
well documented from several reported measurements. The thrust mechanism
in most of these cases was a panoramic Mao-Bell cell with a capacity of approximately 0.5 tonne, in some cases clamped into a load frame which gives
additional force up to 5 tonnes. Such a "moissanite anvil cell" (MAC) is not
very different to a diamond anvil cell and is shown in Figure 7.6. Xu et al. investigated the pressure performance of moissanite anvils as a function of culet
size and gasket dimensions, as well as with and without radial support [66].
Anvils with diameters between 4 and 8 mm were used, having circular culets
with diameters of 0.2-4 mm and a pavilion angle of 45°. Gaskets where made

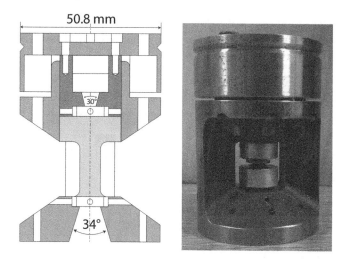

FIGURE 7.6
Panoramic pressure cell used with moissanite and diamond anvils [66]. Drawing and picture provided by M. Guthrie and C.L. Bull.

of T301 steel with starting thicknesses between 0.26 and 1.0 mm, depending on the culet size. To give a representative example, with a 4 mm culet size and an initial gasket thickness of 0.5 mm, a pressure of 25 GPa is reported with a final gasket thickness of 0.3 mm. Sample volumes of 10 mm³ are reported to be compressed to 12 GPa and 0.4 mm³ to 40 GPa. All of these measurements were carried out without pressure transmitting medium. According to these tests, the use of radial support of the anvils provided by binding rings is essential to reach high pressures at large volumes.

Neutron diffraction data using such anvils were reported by Ding et al. [67] where a powder sample of 4-5 mm³ volume was compressed between anvils of 6 mm culet diameter and a TiZr gasket with a hole of 2 mm diameter. There was no pressure transmitting medium and a pressure of 20 GPa was reached. The surprising absence of any observable magnetic reflections might be related to the non-hydrostatic conditions in these measurements. Using a similar set-up, Guthrie et al. carried out measurements on GeO$_2$ using supported anvils of 8 mm diameter and 2 mm culet, with a 9 mm outer diameter T301 gasket with a hole of 1 mm diameter and an initial thickness of 1 mm. A maximum pressure of 16 GPa was reached on a sample volume of approximately 0.8 mm³, again under non-hydrostatic conditions (M. Guthrie, priv. comm.). Tests carried out by C. L. Bull et al. using fretted anvils with 5 mm outer diameter and 4.5 mm height reached a maximal pressure of 7 GPa on a Cu sample without pressure transmitting medium (C.L. Bull, priv. comm.). Unfretted anvils using water

as sample would always fail below 2 GPa, either in the liquid below 0.8 GPa
or at 2 GPa upon heating close to the melting line (approximately 80 °C).

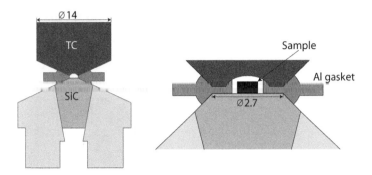

FIGURE 7.7
Hybrid moissanite-TC anvils for single crystal studies at low temperatures,
after Osakabe et al. [68]. Left: overall view, right: close up view of the pressure chamber (schematic). The gasket material is an Al-alloy (JIS A2017P, a
duralumin type alloy). Redrawn from Ref. [68].

Figure 7.7 shows an interesting anvil geometry applied by Osakabe et al.
for single crystal neutron diffraction [68]. The obvious aim was to increase the
sample volume by using "cupped" anvils similar to those applied for sapphire
anvils (Section 7.1.2). In this cell, only one anvil is made of radially supported
moissanite with a flat culet, whereas the other is made of tungsten carbide
with a cup to allow single crystals of 0.5 mm thickness. The flat culet has the
advantage to maintain a defined orientation of the crystal as the pressure is increased. With a gasket made of an Al-alloy and glycerine as quasi-hydrostatic
pressure transmitting medium, pressures of up to 10 GPa were reached on
samples of typically 0.2 mm^3. A similar set-up with unsupported sapphire instead of moissanite reached only 7 GPa, which again underlines the superior
mechanical performance of moissanite. The force-generating device in these
experiments was a panoramic clamp of 45 mm diameter and 80 mm height
made of CuBe, very similar to the devices shown in Figure 7.1. Diffraction
measurements have been carried out at simultaneously 1.5 K, magnetic fields
of 6 T, and pressures of 4 GPa [68].

To resume, from the experience obtained by various groups it is clear
that powder volumes in mm^3 size can be compressed to 10-20 GPa using
supported moissanite gem anvils, certainly under non-hydrostatic conditions,
and possibly even under quasi-hydrostatic conditions. Nevertheless, the claim
of Xu et al. to have reached 40 GPa and beyond on sample volumes of 0.4 mm^3
is surprising and has not been confirmed by other groups. The reason might
be intrinsic to the setup of this experiment. In this case the sample was in fact
a mixture of diamond and ruby powder, which means that the individual ruby

grains were mostly in contact with and compressed by diamond whereas the moissanite was essentially a backing plate. Given the vastly superior strength of diamond, the pressure generation is expected to be beyond what is possible in moissanite anvils using other samples and under more standard conditions.

A possible reason for the limited strength of moissanite in hydrostatic pressure conditions is the presence of visible flaws in the majority of gems, in particular large anvils with diameters of 5 mm and beyond. Figure 7.8 shows typical imperfections which are in the bulk and can extend over a length of several millimeters. Such "tubings" contain often an accumulation of micro-bubbles. If such flaws extend up to the surface they will act as a sink for any low-viscosity fluid and unavoidably lead to anvil failure. The fact that under hydrostatic compression failure occurs almost always through a clean radial cleavage across the diameter of the anvils seems to support this hypothesis. It appears therefore likely that large volume moissanite anvils without such flaws can be taken to considerably higher pressures than possible so far.

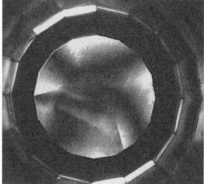

FIGURE 7.8
Typical flaws in large gem anvils: Left: Flaws ("tubings") in a moissanite anvil of 16 mm diameter and 14.2 mm hight, seen through a 5.6 mm culet. Right: Flaws in a diamond anvil of 4.5 mm girdle diameter and a 2.5 mm culet, observed in cross-polarized light. Courtesy of C.L. Bull.

7.3 Diamond anvil cells (DACs)

The use of diamond anvil cells for neutron scattering was pioneered by scientists at the Kurchatov Institute in the 1980s who reached 31 GPa on a 10^{-2} mm^3 single crystal of deuterium and determined its equation of state [69]. More recently a very similar experiment by Goncharenko and Loubeyre extended this study to 38 GPa [70], and low temperatures, and there are un-published reports of the SNAP/GLW group who recorded data at 57 GPa on a 0.06 mm^3 ice VII powder sample using conical diamonds of 1.5 mm culet size (M. Guthrie, priv. comm.). Despite these spectacular results, it is fair to say that the overall impact of diamond anvil cells in high pressure neutron scattering has so far been very limited. But since progress in neutron instrumentation (by the application of focusing optics, for example) will undoubtedly lead to increasing fluxes on the sample, the use of DACs will likely become more popular, and unavoidable if pressures in the Mbar-range are aimed for. We will therefore discuss this issue in more detail.

The obvious drawback of standard DACs is the extremely small sample volume they can provide. As far as diffraction is concerned, there are potentially three ways to overcome this obstacle: (a) the use of isotopically enriched samples with large scattering lengths, (b) the use of single crystals where the signal to background ratio is much higher since the intensity of the Bragg reflection is localized in reciprocal space, and (c) the use of larger diamonds to increase the sample volume. The problem with (a) is that although isotopically enriched samples can be obtained in mg quantities necessary for DAC work to reasonable prices, the proper sample characterization needs in general much larger quantities. Even if there is a suitable and available isotope (see Table 7.3 for a selection on strongly scattering isotopes), the preparation of isotopically enriched samples containing more than three elements is certainly a major obstacle for the general user of neutron facilities. The use of single crystals (b) appears to be a more realistic option, in particular when single crystals are already available at ambient conditions and do not have to be grown in situ in the DAC. But here the obstacle is that most solids undergo first-order phase transitions in the 0-10 GPa range which in most cases destroy or severely degrade the single crystal. There are a number of systems which tend to form single crystals, even once they went through phase transitions, for example molecular solids and alkaline and earth alkaline elements. But for most of the materials of interest in solid state physics and chemistry this is not the case, and investigations of high-pressure phases have to be carried out on powder samples. There is finally option (c), but the use of large single-crystal diamonds has so far been restricted to quite low pressures, see Figure 7.4 and the discussion below. The availability of large nano-polycrystalline diamond (NPD) anvils as mentioned in chapter 4.4 might radically change this situation and open the perspective to carry out neutron scattering on 0.1-1 mm^3 samples up to the several 10 GPa range. This would however need load frames of much higher capacity than standard DACs can provide, i.e. 10-100 tonnes, and hence would reassemble more Paris-Edinburgh type presses.

7.3.1   DAC load frames

The design of load frames adequate for neutron scattering is, in principle, not different to those used for x-ray scattering, and is discussed already in

TABLE 7.3

Isotopes with coherent scattering length b larger than 9 fm.
Units: fm=10^{-15}m, "res" indicates a scattering resonance.
For the complete list of scattering lengths see Appendix G.

Element	b	Element	b	Element	b
^{14}N	9.4	^{76}Se	12.2	^{189}Os	11.0
N	9.3	^{147}Pm	12.6	^{190}Os	11.4
^{35}Cl	11.7	Sm	res	Os	10.7
Cl	9.6	Gd	res	^{192}Os	11.9
^{36}Ar	24.9	^{160}Gd	9.2	Ir	10.0
Sc	12.0	^{161}Dy	10.3	^{194}Pt	10.6
^{58}Fe	15.0	Dy	17.1(res)	Pt	9.50
Fe	9.6	^{164}Dy	49.4	Hg	12.7
^{56}Fe	10.1	^{166}Er	12.3	Pb	9.40
^{58}Ni	15.0	^{168}Er	10.2	^{232}Th	10.0
Ni	10.3	^{170}Er	10.9	^{235}U	9.80
^{65}Cu	11.1	Yb	12.6	Np	10.6

length elsewhere [4]. There are however two features which need the design of a DAC to be adapted to such work: (A) whereas the wavelength of x-rays used for diffraction in DAC work is in general around 0.5 Å, the wavelength for neutrons is rarely below 1 Å, in general 1.5-4 Å. This entails large scattering angles to access a large Q-range (a large number of Bragg reflections) and hence requires cells which are as "panoramic" as possible with the possibility of scattering in the axial (through both anvils) as well as in the radial (through the gasket) directions. (B) The need for intensity entails the use of relatively large anvils and hence the need of relatively large forces, in the order of 1-10 tonnes ($\approx$ 10-100 kN). The load frames are therefore relatively massive.

Figure 7.9 shows Besedin's design which was occasionally also used with sapphire anvils. The load is applied by two sets of bolts, one for pre-loading and a second set for providing larger forces. The diameter of this cell is less than 100 mm, it has a mass of 5 kg, and a usable solid angle of 1.8π. There is another model which provides a solid angle of up to 2.4π. These cells were used with both diamond anvils (0.8-2 carat, culet size 0.7-2 mm) and sapphire anvils (culet size 4 mm) [71].

Figure 7.10 shows Goncharenko's 'hybrid cell' used for single crystal work on deuterium [70]. Here a force of up to $\approx$ 8 tonnes (80 kN) is generated by a combined bolt drive and a gas membrane. A membrane alone, given the restriction of the dimensions imposed by the cryogenic equipment, is hardly able to generate the required forces using a gas pressure of 200 bars provided by standard compressed He bottles. The major part of the force (6 tonnes) is hence produced by the bolt drive, and fine adjustment is done by the membrane drive. The relatively large solid angle in the axial direction (56° and 29°)

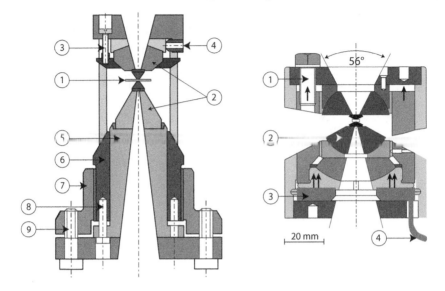

FIGURE 7.9
Diamond anvil cells used for neutron scattering. Left: Besedin's cell [71]. (1) anvils and gasket; (2) support; (3-4) alignment screws; (5) piston; (6) cylinder; (7) heavy load flanges; (8) pre-alignment bolts; (9) heavy load bolts. Right: Goncharenko's type 1 hybrid cell [72]. (1) Pre-loading bolts; (2) support (cBN); (3) membrane; (4) capillary. Arrows indicate forces generated by the bolts (↑) and the membrane (↑↑). Adapted from refs. [71] and [72].

allows the use for synchrotron x-diffraction using image plates which can be crucial for the characterization of the sample prior to the neutron experiments.

7.3.2   Diamond anvils

The lack of intensity in DAC experiments could, in theory, be partly compensated by the use of larger diamonds. There is sufficient documented experimental evidence that such a 'naive' approach has severe limitations, i.e. a simple scaling of the sample volume by the use of larger diamonds (and forces) and at the same time maintaining the pressure performance is unrealistic. As already mentioned earlier (section 7.1.2 and Figure 7.4), the maximum achievable pressure decreases strongly with the size of the anvil, hence the sample volume. This is almost certainly due to flaws which are unavoidable in large anvils as soon as the diameter exceeds typically 2 mm, see Figure 7.8. Bull et al. (priv. comm.) carried out test on a dozen of natural and synthetic (high P/T grown) gems with girdle diameter between 4.5 and 6.0 mm and a culet size between 2.5 to 4.0 mm. Whereas some of these anvils could be taken to 5-7 GPa using solid samples of 0.2 mm^3 volume and no pressure transmitting

FIGURE 7.10
Goncharenko's hybrid type 1 cell disassembled. Reprinted from ref. [72].

media, they systematically failed at pressures below 2 GPa when liquids were used, and by heating to moderate temperatures of up to 200°C. This observation and the way the fractures occurred indicates that it is not due to wrong gasketing or an inadequate seating. It does not depend either on the gem cut since Boehler-Almax type anvils of this size show the same limits in strength. Such problems disappear when the size of the anvils is reduced. Glazkov et al. used in their 31 GPa run on deuterium a culet size of only 0.7 mm diameter [71], very similar to Goncharenko and Loubeyre who reached 38 GPa with a 0.6 mm culet size [72]. In all cases the sample volume was not larger than 10^{-2} mm^3 at the start of the experiment.

To conclude, with the gems presently available it seems to be difficult, if not impossible, to achieve pressures beyond 50 GPa with sample volumes larger than 10^{-2} mm^3. This might change with progress in the development of either flaw-free gems or poly-nanocrystalline diamond.

8

Special designs

CONTENTS

8.1 SANS high pressure cells

The difficulty of small-angle neutron scattering (SANS) is that diffraction is recorded at very small angles which means that shielding of the pressure cell material hit by the neutron beam is virtually impossible. In addition, the signal from standard pressure cell alloys interferes in general strongly with the sample signal which makes the substraction of the background difficult. A large field of applications for SANS under pressure concerns research on soft condensed matter where the sample is liquid and the relevant pressure range a few 0.1 GPa. For such purposes there are a number of solutions which are conceptually very similar. The key ingredient in these cells is the use of windows made of single crystalline material. The only material which has sufficient strength and which can be obtained in large sizes is sapphire. The pressure cell by Gabke et al. [73] shown in Figure 8.1 is one of the simplest designs and illustrates the basic principle of such cells. Two sapphire windows enclose a sample volume in the form of a disk. The sample is injected into the pressure chamber by a hole from the side and remains in permanent contact with an external pressure manometer and a hydraulic compressor. The sapphires are kept into their position by two closure nuts which have central bores for the incoming and diffracted beam. There is the possibility to change the temperature by circulating water of the desired temperature around the cell. Earlier versions with similar design were presented by Vennemann et al. [74] (max. pressure 0.2 GPa), Takano et al. [75] (max. pressure 0.2 GPa and temperatures up to 520 K), as well as Pépy and Baroni [76] (max. pressure 0.15 GPa) and Bonetti and Calmettes [77] (max. pressure 0.3 GPa).

A slightly more sophisticated pressure cell of this type is shown in Figure 8.2 [78]. This cell is more massive since it uses considerably larger sapphire windows of 62 mm diameter and 35 mm thickness and has increased pressure performance with a maximum pressure of 0.5 GPa. The available sample

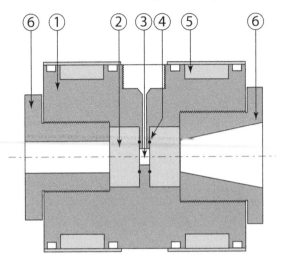

FIGURE 8.1
High pressure cell for small-angle neutron scattering (SANS) to 0.3 GPa, after
Gabke et al. [73]. (1) body, (2) sapphire anvils (Ø10 mm), (3) sample chamber,
(4) O-ring seal, (5) water circuit, (6) closure nut. Redrawn from Ref. [73].

volume is 1-1.5 cm^3, hence two orders of magnitude larger than Gabke's and
Takeno's design. The hydraulic fluid used by the compressor is separated from
the sample fluid by a Viton O-ring which is sufficiently elastic to transmit the
pressure without loss. This elegant and simple method was already applied in
Takeno et al.'s SANS cell [75]. The O-ring is part of the sample "pill" which
can be pre-loaded and then inserted into the SANS cell at the time of the
measurement.

SANS cells of this type allow the possibility of time resolved diffraction
using the "pressure jump technique" to study the dynamics of phase transition. For this purpose the pressure is increased abruptly within a second or
less and the evolution of the SANS signal is observed as a function of time,
typically over hours. Another interesting aspect of such pressure cells is that
the optical transparency of the anvils allow light scattering to be performed
simultaneously.

An important feature of SANS high pressure cells is that the sample thickness along the beam needs to be as stable as possible to obtain reliable intensities. In most cases this is not possible by mechanical means and the variation
of the sample thickness with pressure needs to be calibrated and the data
corrected for absorption accordingly.

A slightly different strategy - but still based on two opposed sapphire single
crystals - is Bonetti and Calmettes' 2005 design [77]. Here the sapphires are no
longer simple windows but anvils with culets which deform a metallic gasket

to generate pressure. This cell should hence be classified as sapphire opposed anvil cell and is indeed very similar to the concepts presented in chapter 7. The advantage seems to be the increased pressure performance (0.53 GPa), but pressure jump experiments are clearly excluded, and the pressure dependence of the sample thickness is appreciable and needs to be corrected for.

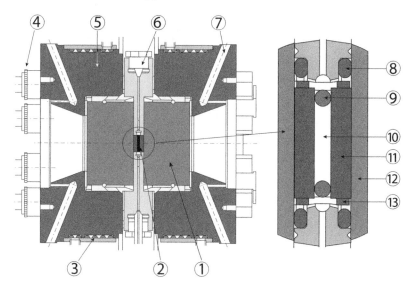

FIGURE 8.2
High pressure cell for small-angle neutron scattering (SANS) to 0.5 GPa [78]. (1) Sapphire window; (2) sample cell; (3) cooling jacket; (4) tie rods; (5) cell body; (6) hydraulic oil inlet; (7) heater holes; (8) metal seal; (9) Viton ring; (10) sample chamber; (11) sapphire disc; (12) sapphire window; (13) brass ring with holes. Redrawn from Ref. [78].

8.2 Single-crystal sapphire gas cells

The use of liquid and gas cells made of metal as discussed in chapter 5.1 has a number of inconveniences, in particular when used on continuous neutron sources. All metals produce Bragg reflections or incoherent background which are difficult to shield at low and high diffraction angles. Most of these materials absorb neutron and are incompatible with hydrogen, in particular at high temperatures. For these reasons there have been a few attempts to construct gas/liquid cells made of single crystalline sapphire, see Figure 8.3. The advantage of sapphire is its very low neutron absorption (see chapter 7) and its stability under contact with hydrogen, even at elevated temperatures. The advantage of single crystals compared to sintered Al_2O_3 is, a priori, the possibility to obtain an extremely weak background with only a few isolated Bragg reflections from the pressure cell.

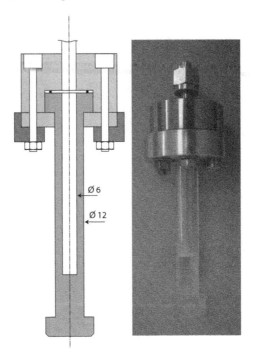

FIGURE 8.3
Gas pressure cell made of single-crystalline sapphire, after Kohlmann et al. [79]. See also Rondinone et al. [80] for a similar design for pressures up to 38 MPa.

Possibly the first design of such a sapphire gas cell is reported by Rondinone et al. [80] who used it for inelastic powder neutron scattering in the 0-30 meV energy transfer range investigating THF-clathrates at low temperatures. The inner and outer wall diameters were 10 and 24 mm with end flanges as in Figure 8.3 which shows a similar but more recent design by Kohlmann et al. [79]. The maximum pressure reached in Rondinone's device was 38 MPa (ca. 380 bar), determined not by the crushing of the sapphire but the failure of the indium seals. Using formula 1.19 and a tensile strength of 400 MPa [80, 63], one would expect a pressure limit far beyond, i.e. approximately 2 kbar. The extremely small fracture toughness of sapphire (Table 7.2) requires however particular care in the machining of the cell, in particular a careful polishing of the inner wall surfaces.

Figure 8.4 shows a map of diffraction patterns of an empty sapphire cell oriented with the c-axis vertically, at different cell orientation angles ω [79]. They reveal that there is considerable thermal diffuse scattering which produces contaminant background over a large 2θ range. Fortunately the scattering is highly anisotropic which allows to avoid it by choosing a certain cell orientation, indicated by the arrow in Figure 8.4. With this setting, the contamination from thermal diffuse scattering close to the nearby (300) Bragg reflection is only over a range of approximately $5°$.

This pressure cell is routinely used for time-resolved diffraction experiments investigating the formation of metal hydrides in contact with hydrogen/deuterium gas under high pressure and high temperature to 16 MPa and 700 K [79]. The sapphire allows remote laser heating of the sample and temperature measurement by pyrometry. In this way structural parameters can be obtained by Rietveld analysis with data collection times of one minute or less.

FIGURE 8.4
Background produced by the single-crystal sapphire cell shown in Figure 8.3, illustrated as a map of diffraction patterns at different cell rotation angles ω. Data were obtained at $\lambda=1.87$ Å at the D20 diffractometer of the ILL. The indices (hkl) correspond to Bragg reflections of sapphire, the diffuse scattering is due to phonon excitations. The arrow at $\omega=20°$ would be a favorable setting for measurements with low background. Courtesy of H. Kohlmann.

9

Uniaxial pressure cells

CONTENTS

9.1 General

The use of uniaxial pressure is a unique way for investigating the anisotropy of interatomic interactions. What ideally is desired is a method which changes only one lattice parameter while keeping the others strictly constant. Such a uniaxial *strain* experiment is unfortunately difficult, if not impossible to carry out practically. What is instead feasible is a uniaxial *stress* apparatus, where a single crystal is subjected to a force along a certain crystallographic axis. For the purpose of simplicity let us take the case of an isotropic sample where the elastic response can be described by only two parameters, Young's modulus E and Poisson's number ν. Given a certain force F along x on a surface S, the stress-strain relations are given by:

$$\sigma_1 = -F/S \qquad (9.1)$$

$$\sigma_2 = \sigma_3 = 0 \qquad (9.2)$$

$$\epsilon_1 = \sigma_1/E \qquad (9.3)$$

$$\epsilon_2 = \epsilon_3 = -\epsilon_1\nu \qquad (9.4)$$

$$dV/V = \epsilon_1(1 - 2\nu) \qquad (9.5)$$

Since ν is usually positive, the effect of uniaxial stress is to shorten the sample along the compression axis and to expand it perpendicular to it. For a single crystal the stress-strain relations are more complicated and depend on the crystal structure, but qualitatively the behavior is similar. Carrying out several uniaxial pressure experiment with the compression along different crystal axes as well as hydrostatic measurement will, in general allow a conclusion on the anisotropy of the interaction under investigation.

The considerable disadvantage of uniaxial pressure experiments is the very limited pressure range. To illustrate this let us take again the case of an isotropic sample. According to Tresca's criterion, plastic deformation occurs whenever $|\sigma_1 - \sigma_3| \geq 2\tau$ where τ is the shear strength. This means (equations 9.1 and 9.2) that for uniaxial compression along x on a sample surface S, plastic deformation occurs for a force $F_{max} = 2S\tau$. Experience shows that in the case of a single crystal the related pressures (stresses) are typically a few kbar or less.

The use of uniaxial pressure devices in neutron scattering was enormously promoted by the interest in layered magnetic systems as well as high-Tc superconductors. Since the samples under investigation are single crystals, the volume can be as small as a few mm^3 which allows the use of relatively small devices which can at the same time be used for magnetic measurements under high fields.

9.2   Clamp devices

Figure 9.1 shows a uniaxial pressure cell with 40 mm outer diameter by Aso et al. [81] designed for measurements to 10 K on a four circle diffractometer at the Japanese research reactor JRR-3M. Most parts of the cell are made of 7075 type Al-alloy (chapter 2) which causes negligible neutron absorption and allows access to the entire equatorial plane of the sample. A force of up to $\approx$ 2 tonnes (20 kN) is applied on the upper backing disc using a tool (not shown) and locked by the upper lock nut. The clamp can be equipped with two types of disc springs (CuBe and steel) which allow adjustment of the load to the elastic properties of the sample. The piston is made of several ZrO_2 cylinders with carefully polished and highly parallel surfaces and allows sample dimensions of typically 3 x 3 x 1 mm. Pressures of up to 0.4 GPa are reported in diffraction measurement on various metal oxide samples [81].

Figure 9.2 shows two types of uniaxial pressure cells which can be used both for neutron scattering and susceptibility measurements under high magnetic fields due to their relatively small diameter of 12 mm [82]. The two cells differ in the direction of compression relative to the cell axis but are otherwise identical. A thrust of up to 20 kN is generated by the top screw and transferred to a barrel spring which maintains the force during temperature changes. The barrel springs transfer the force further to a spacer and then to two zirconia anvils. In the L-type ("longitudinal") cell the anvils have a Bridgman geometry, i.e. a conical front face with a flat tip of 4 mm diameter. In the T-type ("transverse") cell the force is deflected by 90° by two guiding blocks (very similar to certain multi-anvil devices) and then applied to the zirconia anvils. In both cases the faces in contact with the sample are polished and have a planarity of typically less than 2 μm. All metallic parts of the cells are made of non-magnetic TiAl6V4 alloy (Table 2.4), an earlier version used CuBe and Belleville washers instead of barrel springs [82]. The single crystal

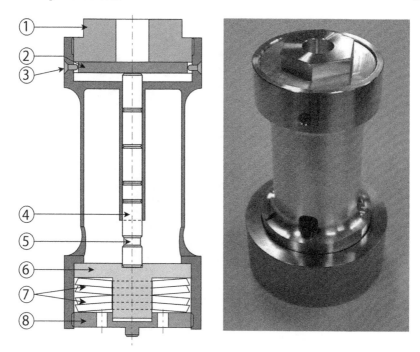

FIGURE 9.1

Uniaxial pressure cells after Aso et al. [81]. (1) lock nut (upper); (2) backing disc; (3) anti-twist screw; (4) piston (zirconia ZrO_2); (5) sample; (6) lower backing disc; (7) disc springs (steel); (8) lock nut (lower). If not specified otherwise, items are made of Al-alloy 7075. Photograph courtesy of Dr. N. Aso.

samples need to be carefully polished and need to have highly parallel faces. Zirconia (in short ZrO_2) has a shear strength of 1.5 GPa which means that, in principle, single crystals can be compressed over a wide range of stress given by $\sigma \leq 2\tau_{sample} \leq 2\tau_{ZrO_2}$. The stress onto the sample is determined simply by the applied force onto the sample, which can be calculated from the change in the length of the barrel spring and a calibration curve. The compression of the barrel spring can be measured by the position of the top screw with respect to the edge of the body.

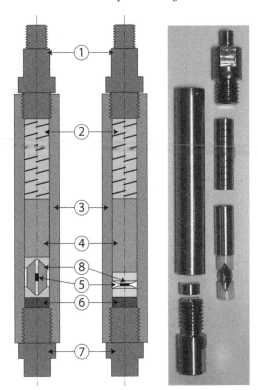

FIGURE 9.2

Uniaxial pressure cells after Kamarád et al. [82]. Middle: "Longitudinal" (L)
cell for compression along the cell axis. Left: "Transverse" (T) cell for compression perpendicular to the cell axis. (1) compression screw; (2) barrel springs;
(3) cell body; (4) and (6) pistons; (5) sample; (7) end nut; (8) anvils. Anvils
(8) are made of zirconia, all other parts are made of TiAl6V4. Note that the
length/diameter ratio has been reduced in the drawings. Photograph courtesy
of Dr. J. Kamarád.

9.3   Externally driven devices

These are devices where the force on the sample can be transmitted and
adjusted in situ using some kind of push-rod and an external force generating
mechanism attached to it.

Figure 9.3 shows such a type of uni-axial cell used by Hälg et al. for
low-temperature magnetic measurements [83]. The cell itself is very similar
to Aso's design (Figure 9.1) with a body of 30 mm outer diameter made of
high-strength aluminium alloy. The pressure can be changed in situ, at low

temperatures, using a manually driven hydraulic ram which transfers a force of up to 10 kN via two concentric hollow tubes onto the sample. Anvils with strictly flat and parallel culets were made of stainless steel, aluminium, or quartz, depending on the nature of the sample. The tip of the pushrod is rounded which allows for some self-alignment of the upper anvil it pushes on. The cell can be inserted into a standard orange cryostat with 50 mm bore to allow measurements down to 2 K.

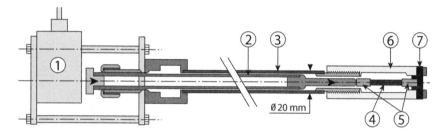

FIGURE 9.3
Uniaxial pressure cell for magnetic measurements at low temperatures, after Hälg et al. [83]. (1) hydraulic ram; (2) inner (force transfer) tube; (3) outer (force retaining) tube; (4) sample; (5) anvils; (6): cell body; (7): bottom plate. (2),(3) and (5) are made of stainless steel, aluminum or quartz, (6) of an Al-alloy. The length of the tubes is approximately 1 m. Adapted from ref. [83].

A low-temperature uni-axial pressure cell for cm-size samples was used by Krystian et al. [84] for small angle neutron scattering (SANS), see also chapter 8.1 on this subject), see Figure 9.4. The set-up was originally used for small angle x-ray diffraction. A force of up to 10 kN is generated by an electric deformation drive and transferred to the sample chamber by a quartz tube which sits on top of the upper piston. Contrary to the setup of Hälg et al.'s cell (Figure 9.3), cooling is not achieved by a cryostat but by a closed cycle refrigerator which needs to be placed below the sample and thermally coupled to it by flexible copper strips. Fast cooling is possible by an additional liquid nitrogen reservoir. With this device pressures can be changed in situ in either a constant strain or constant stress mode, at temperatures down to 5 K. The SANS measurements required only diffraction in the forward direction and hence only a limited aperture for the diffracted beam. It should be possible to modify this set up to allow also wide-angle diffraction.

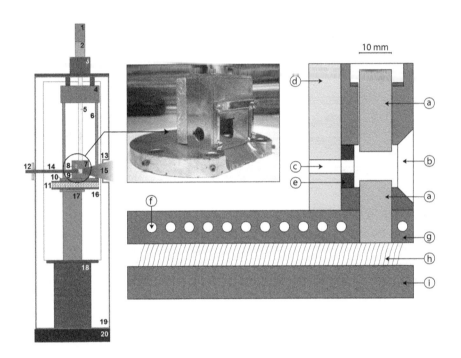

FIGURE 9.4
Uniaxial pressure cell for SANS measurements by Krystian et al. [84].
Overview (left) and enlarged view of the sample chamber (right). (1) Motor; (2) gears; (3) force transducers; (4) (liquid N_2 container; (5) quartz tube;
(6) stainless steel tube (4 in total); (7) sample chamber (removable); (8) temperature sensor (Si-diode); (9) sample position; (10) heater; (11) copper wires;
(12) window (fused silica); (13) window (100x100 mm); (14) incident beam;
(15) diffracted beam; (16) thermal radiation shield; (17) cold head - second
stage; (18) cold head - first stage; (19) cryostat wall; (20) cryostat base; (a) piston; (b) window (diffracted beam); (c) window (incident beam); (d)-(e) front
collimators; (f) heaters; (g) Cu-plate; (h) Cu wires; (i) Cu-plate in contact
with low-T head. Photograph courtesy of Dr. W. Pichl.

10

Paris-Edinburgh cells I

CONTENTS

10.1   General

What is nowadays termed *Paris-Edinburgh cells* (PECs), or *Paris-Edinburgh presses*, are compact hydraulic presses with capacities (=maximal forces) between 50 and 500 tonnes which compress samples of typically 1-100 mm^3 volume using in general opposed anvils made of some sintered materials (tungsten carbide, diamond, boron nitride). The name refers hence to the type of load frame, not to the type of anvils, very much in the spirit of the "Merrill-Basset", "Mao-Bell", and "Syassen-Holzapfel" cells used in diamond anvil technology. "Compact" means in this context more or less portable, i.e. the presses have masses of only 10-100 kg, despite their considerable capacities. Paris-Edinburgh cells were developed in the early 1990s in a program initiated by Nelmes (University of Edinburgh) and Besson (University of Paris) for neutron scattering in the 0-10 GPa range on spallation sources, at that time the U.K. ISIS facility [85]. The technology has spread meanwhile over four continents and has found high pressure applications beyond neutron scattering. Since PECs are distributed by a sole company with strong links to the groups who initiated the program, the pool of pressure cells is homogenous and compatible worldwide. This provides an enormous advantage for users who tend to work on multiple neutron sources and who carry high pressure equipment between them. The impact of PECs on the landscape of high pressure neutron scattering cannot be underestimated, and it is likely that this technique will continue to play a major role in the future, even on medium-intensity sources. A large part of this book will hence be devoted to this subject.

10.2   Load-frames

Paris-Edinburgh presses can be divided into two types as illustrated in Figure 10.1: the original four-column V-design [85] (from French verin = ram) and the two-column VX-type which came available from 2003 on [86]. A compact two-column press was also proposed by Stishov and Sadkov [87]. The choice of one or the other geometry is very much dictated by the type of neutron source where the cell is used as well as its applications: pulsed sources collect data usually in an energy-dispersive mode with a constant diffraction angle 2θ where the presence of four columns is not an inconvenience. In contrast, continuous sources need large contiguous windows to record data over a large 2θ-range which can only be provided by the VX-type cells with their two horizontal openings of 160° each.

The two types of presses shown in Figure 10.1 exist in various sizes, i.e. capacities, depending on the force required for the particular applications. Whereas the smallest press (VX1) has a capacity of 50 tonnes and a mass of less than 10 kg, the largest cell constructed so far (V7) has a capacity of 450 tonnes and a mass of 95 kg. Table 10.1 gives a complete list of the different types and sizes so far available along with their principal characteristics. Apart from force, available space around the sample is a major criterium which might determine the choice of the type of cell.

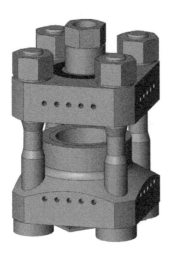 

FIGURE 10.1
PE-load frames with a capacity of 130 tn. Left: Four-column V-type cell; right: Two-column VX-type.

TABLE 10.1

Types of PE presses and principal characteristics. The length of the tie rods can be adapted to particular applications. Capacity = max. load in tonnes. Diameter = max. dimension across diagonal. All dimensions are in mm.

type	capacity	mass (kg)	columns	rear access	diam. ram	diam. piston	height
VX1	50	10	2	yes	120	50	188
VX2	50	10	2	yes	120	50	188
V3	250	50	4	no	248	114	309
VX3	200	50	2	yes	230	114	315
V4	250	50	4	yes	248	114	309
VX4	200	50	2	yes	230	114	315
V5	150	35	4	no	198	92	263
VX5	130	35	2	no	180	92	242
V7	450	90	4	no	305	150	412
V8	450	90	4	yes	305	150	412

A key feature of PE presses is their "compactness", i.e. their high capacity compared to their mass. A V3 (VX3) cell has a capacity of 250 tonnes but only a mass of 50 kg. Commercially available presses of this capacity have masses approximately one order of magnitude larger. This has been made possible by applying finite-element methods in the design to optimize the weight, as well as by the use of high tensile steels.

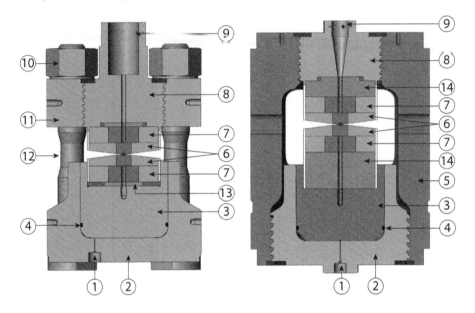

FIGURE 10.2
Cross section of Paris-Edinburgh presses type V (left) and type VX (right). (1) Hydraulic fluid inlet; (2) cylinder; (3) piston; (4) O-ring seal; (5) load frame; (6) anvils; (7) TC backing plates (seats); (8) breech; (9) front collimator; (10) nut; (11) top platen; (12) tie rod; (13) backing disc; (14) steel spacer.

Figure 10.2 shows cross sections of the two types of cells including the anvil assembly. The characteristic feature of all PE presses is the unusual shape of the ram, in particular the cylinder which will be discussed further below. A hydraulic fluid is injected through an inlet (1) and pushes on a piston (3) with cross section A. For the most common cell, the V3/VX3 type cells, A=100 cm^2, i.e. a fluid pressure of 1 kbar generates a force of 100 tonnes on the piston. Such forces cannot be transferred directly onto the anvils (6) since the contact surface with anvil die (typically 3-5 cm^2) is not sufficient to prevent indentation of the steel which has a yield strength of maximal 15 tn/cm^2. The anvils are therefore seated by steel-supported tungsten carbide (TC) or diamond backing plates (7). The diameter of the dies is typically 40 mm and the contact surface with the steel hence 12 cm^2 which indicates a maximal allowable load of 170 tonnes which is sufficient for the majority of experiments.

If the required load is beyond this value, backing discs with larger dies have to be used.

The upper part of the cell consists of a breech (8) which is screwed into the top platen (11) or the top of the load frame (5). Attached to the breech is the upper anvil-backing disc assembly which allows a very quick and easy removal by simply unscrewing the breech. This design has the additional advantage that slight differences in the thickness of the gasket and the anvils can easily be compensated by adjustments on the position of the breech. In fact, experience shows that even if the breech is unscrewed by 5 mm relative to its ideal position, the press remains fully operational up to its maximum load.

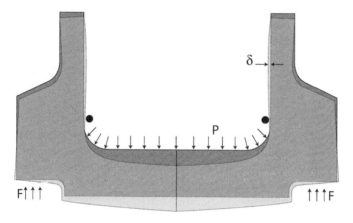

FIGURE 10.3
Deformation of a PE ram under load derived from finite-element calculations. A hydrostatic pressure P is applied on the bottom of the cylinder up to the level of the O-ring seal ($\bullet$) and the counter-forces F act on the area indicated by vertical arrows ($\uparrow\uparrow\uparrow$). Note the absence of radial deformations close to the O-ring seal. Deformations are strongly exaggerated to make them more visible. Maximal radial deformations δ for a VX3 cell are 0.05 mm.

A characteristic and critical ingredient of PE presses is the shape of the cylinder which deforms under load in a particular way and thereby insures leak tightness. Figure 10.3 illustrates results of finite element calculations of an undeformed cylinder and a cylinder subjected to a pressure of 250 MPa on the inside, up to the level of the O-ring seal. The calculations assume rotational symmetry, and that the counterforces of 250 tonnes act on the lower part, at the position of the four tie rods. It is seen that the deformations of the cylinder are considerable, for example the bottom bends by almost 0.5 mm. However, the radial deformations of the inside are almost zero, in particular at the level of the O-ring where the radial displacements are less than 0.05 mm, hence less than the tolerance between cylinder and piston. In other words, the cylinder deforms in such a way that the piston never jams (the deformations of the piston can in fact be neglected). This deformation pattern does not change fundamentally when the real 3D geometry with the four (two) tie rods is considered, though the radial displacements then depend obviously on the direction.

The hydraulic ram shown in Figure 10.2 does not allow access to the rear anvil, neither does it allow transmission measurements. This has a number of disadvantages, in particular for high P/T applications which will be discussed later. The ram shown in Figure 10.4 (lower) removes this inconvenience. It provides a free channel of 3-0 mm diameter (depending on the type of cell) from the back up to the anvil. Note that this design needs a second O-ring seal. The main disadvantage appears for low-temperature applications, in particular if helium is used as hydraulic fluid. The reason is that in this case the simple O-ring seal has to be abandoned in favor of a more complex unsupported area ("Bridgman-") seal, which is difficult to implement for a ram as shown in Figure 10.4 (lower). A given load frame can be fitted with both types of rams and the same anvils and backing disc assemblies.

Given these various options on the capacity and the geometry of the rams, the nomenclature of cells is as following: Cells equipped with rams *without* rear access carry *uneven* numbers, such *with* rear access *even* numbers. A V3 is hence a four-column model with a standard ram (no rear access), whereas a V4 is equipped with a ram with rear access. The same applies for the two-column VX series. As for the number, it refers simply to the chronological sequence of development.

All PE load frames are made of a high tensile steel used for aeronautic applications. It is produced by Aubert & Duval (France) and commercialized under the trade name 819AW and belongs to the class of "alloyed steels" as discussed in chapter 3. The complete absence of cobalt is fortuitous for neutron applications since it avoids the potential risk of activation by the production of ^{60}Co. This steel has been quenched from 875°C to air, cooled to -80°C to remove residual austenite, and then tempered twice to 300°C which produces a yield strength of 1.4 GPa and a tensile strength (UTS) of 1.7 GPa, see Figure 3.2. Most billets used for machining PE presses have been forged prior to the tempering to obtain material in useful starting dimensions. The forging improves the microstructure and is hence beneficial for strength. Given the thermal history of the steel, a temperature of 300°C is the upper limit to which the press can be heated without irreversible reduction in its performance.

The 819AW steel has proven its reliability over the last two decades on approximately 90 PE presses in operation worldwide. It has revealed its out-standing properties in particular in the use at very low temperatures (to below 4 K) where all steels unavoidably tend to embrittle, i.e. where the toughness reduces substantially so that minor defects in the steel can easily lead to fracture. Changing the type of steel used for the construction of PE load frames appears currently not a recommendable option. So far the only exception is the VX1 press which was built entirely of CuBe and which is currently in use at the ILL.

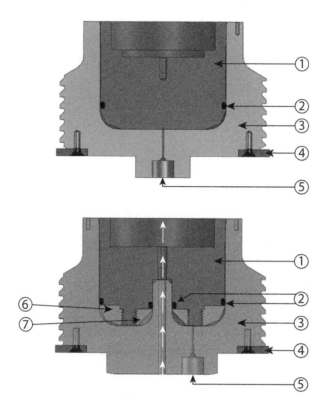

FIGURE 10.4
Cross section of hydraulic rams with (lower) and without (upper) rear access.
(1) piston; (2) inner and outer O-ring seal; (3) cylinder; (4) centering ring; (5)
high pressure fluid inlet; (6)-(7) seal retaining rings. The white arrows indicate
the access for neutrons through the ram.

10.3 Anvils and gaskets

Whereas the mechanics of the load frame can be solved by more or less straight
forward engineering, the problem of anvils and gaskets is much more complex.
The reason is that these parts operate under severe conditions of pressure and
shear, involving large plastic deformations with considerable friction occurring
between them. These conditions are difficult to model and most of the critical
input parameters are unknown. As a consequence, the mechanics of anvils
and gaskets still relies on recipes and experience obtained by a trial-and-error
approach.

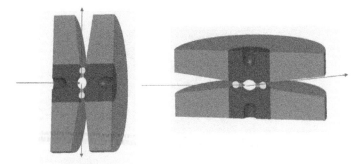

FIGURE 10.5
Scattering geometries used with anvils in PE type cells. Left: *Transverse* scattering (in through the anvil - out through the gasket), and *longitudinal* scattering (in through the gasket - out through the gasket).

With the exception of some high P/T applications discussed in the next chapter, the majority of neutron studies with PE presses use toroidal anvils with central cups. These types of anvils can be classified as *profiled Bridgman anvils*. The cup is important since it increases the sample volume (hence the scattered intensity), reduces the visible background from the anvils, and provides space to accommodate elements for internal heating in high P/T neutron scattering measurements. Bridgman anvils with central cups were used already in the 1950s by Hall who claims that even Bridgman experimented with it [88]. The addition of a toroidal gasket (ring) is an invention of Khvostantsev et al. [4, 89, 90] who used such anvils for physical property measurements as well as material synthesis. Besson realized around 1988 that such anvils are useful for neutron scattering and thereby contributed to their popularity for other applications, for examples synchrotron x-ray scattering.

Toroidal anvils can be divided in single and double-toroidal profiles (in principle also multi-toroidal). The essential difference for neutron scattering is in the pressure performance: whereas the maximal pressure achievable with single toroidal anvils is approximately 13 GPa, double-toroidal anvils are able to reach routinely pressures up to 25 GPa on sample volumes approximately two times smaller than used for single-toroidal profiles. Although there is a priori a considerable choice in dimensioning such anvils, in practice there are only a few geometries which are currently used with the load frames shown in Figure 10.1 and which are therefore described in more detail in the following. They are used for the majority of measurements on powders at ambient and low temperatures. High P/T experiments using internal heating require a different profile and will be discusses separately in section 11.3.2. Also, single-crystal studies require some adaptation of the standard profile. Reference [91] and Appendix D give examples of anvil profiles used for single-crystal diffraction and phonon dispersion studies.

Very similar to diamond anvil cell techniques, the opposed anvils used in PE-type cells can be used in two types of scattering geometries, transverse and longitudinal mode, see Figure 10.5. Transverse scattering imposes a fixed diffraction angle close to 90° is usually applied in pulsed sources whereas the longitudinal mode is adapted to continuous (in general reactor) sources. Both have distinct advantages and disadvantages. To cite a few: in the T-mode the incident polychromatic beam has to cross some 6 mm of anvil material and is thereby strongly distorted by the wavelength dependent absorption (see chapter 4) which has to be corrected if reliable intensities are needed. Its uncontestable advantage is that the shielding of the background is very efficient. It is the ideal geometry for high pressure-high temperature diffraction by internal heating. The same measurements in the longitudinal mode would give strongly contaminated diffraction patterns.

10.3.1 Single-toroidal anvils

Figure 10.6 shows the setup used in the majority of powder neutron scattering (diffraction and inelastic) experiments to 10 GPa, i.e. about 90% of all high pressure neutron scattering experiments carried out currently. It consists of a single-toroidal anvil profile[1] (in contrast to double-toroidal profiles for pressures beyond 10 GPa) with a sample chamber of 6 mm diameter surrounded by a set of metallic gaskets, in most cases null-scattering TiZr (chapter 2.2). The gasket assembly consists of two parts: (a) an outer part in form of a toroid (a ring) which is preformed after machining, and (b) an inner part which can either be a simple washer of 6 mm inner and 10 mm outer diameter (and an initial thickness of typically 1.6 mm, depending on the sample compressibility), or a pair of hemispherical dishes which allow the sample to be completely encapsulated, i.e. without touching the anvils. The encapsulation of samples by such gaskets is an important technical improvement since it allows hydrostatic experiments using the 4:1 methanol-ethanol mixture [92]. We will call them *encapsulating gaskets* throughout the book. Experience shows that whenever a fluid sample (producing hydrostatic conditions) is in direct contact with anvils of whatever type (TC, sintered diamond, boron nitride) the maximum attainable pressure is 3-4 GPa where an anvil failure occurs. Similar anvil failures are often observed when samples under pressure undergo phase transitions. Generally speaking, the lifetime of anvils can be considerably increased by the use of encapsulating gaskets. The small reduction of sample volume can be tolerated in most cases. For solid samples with high shear strength such as oxides which are compressed without pressure transmitting fluid, the use of washer-type gaskets bears usually no risk for the anvils.

The initial thickness of the various gaskets is critical and merits a few

---

[1] For some reason this profile is also called 'Los Alamos profile.'

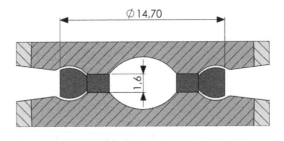

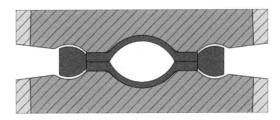

FIGURE 10.6
Gasket assembly in single-toroidal anvils using washer-type (upper) and encapsulating (lower) inner gaskets. Dimensions of the anvil profile are given in Appendix D.

remarks. First of all, there is the general tendency to increase the thickness to accommodate more sample and hence improve the data quality. The room for manoeuvre is unfortunately very limited, and the reason is very simple: increasing the initial gasket thickness leads eventually to the geometry of a simple cylinder which is known to sustain pressures of only a few GPa (chapter 1.2). For the 'Los Alamos' anvil profile shown in Figure 10.6, experience shows that the upper limit of the inner gasket thickness is close to 2.0 mm, unless highly compressible samples are used where the initial thickness rapidly decreases under load. Thicker gaskets used with ordinary samples will lead to a pressure saturation, large extrusion, and in most cases to a blow-out. On the other hand the gasket cannot be too thin since otherwise the attainable volume reduction $\Delta V/V$ and hence the maximum pressure is limited. A value of 1.6 mm is a typical starting thickness. The thickness of the outer (toroidal) gasket is chosen such that it is not deformed in the initial stage of compression, i.e. the force is initially only applied to the inner gasket at a diameter of 10 mm. It is only when the forces exceed typically 10 tonnes that it has sufficiently deformed that the outer gasket starts to work. To give an example, for an inner gasket thickness of 1.6 mm the toroid is typically 2.5 mm thick for the profile given in Figure 10.6. This procedure leads to a high pressure efficiency since the amount of extrudable gasket material is reduced, and hence the overall diameter of gasket material being compressed is minimized. Such variable-thickness gaskets can be easily made on any lathe without any particular tools.

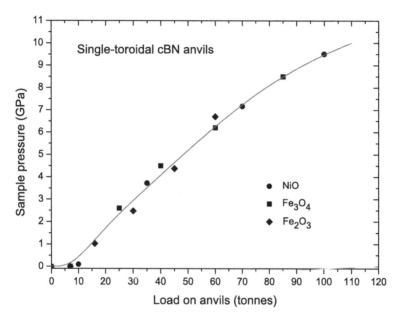

FIGURE 10.7
Typical pressure-load relation for single-toroidal cBN anvils with a profile as shown in Fig. 10.6. Encapsulating TiZr gaskets were used with a 4:1 methanol-ethanol pressure transmitting medium. The samples were NiO, Fe_3O_4, and Fe_2O_3 which served also as pressure markers. The line is a guide to the eye. See text for details.

Figure 10.7 resumes the typical pressure-load efficiency of such a gasket assembly. This calibration curve is based on measurements carried out using cBN anvils, encapsulating TiZr gaskets, and three different powder samples (NiO [93], Fe_3O_4 [94], and Fe_2O_3 (S.K., unpublished)) which served at the same time as pressure markers, as well as a 4:1 ME pressure transmitting fluid.

There is an initial threshold of typically 5-10 tonnes up to which no pressure increase is observed. This corresponds to the minimal load necessary to deform the gasket plastically, given by σ_Y x A where σ_Y is the yield strength of the metal and A the contact area. With the dimensions of Figure 10.6 and σ_Y=0.7 GPa one obtains indeed a force of approximately 10 tonnes. The increase in pressure is then almost linear up to a load of 80 tonnes where the increasing area and the decreasing thickness of the gasket poses increasing resistance to extrusion. For loads beyond 100 tonnes the $P(F)$ curve tends to flatten, mostly because the gasket has become very thin (i.e. 0.2-0.3 mm) and cannot be deformed further, but also because at these pressures the anvil material deforms substantially. The latter effect is less pronounced for sintered diamond which seems to be the reason why the flattening occurs only at higher loads.

Experience with SD anvils with the 'Los Alamos' profile by the Edinburgh group (C.L. Bull, priv. comm.) suggest that up to 8-10 GPa there is no significant difference in pressure efficiency between cBN and SD anvils. For pressures beyond 10 GPa, SD anvils perform better since they can support considerably higher loads. The current experience obtained at PSI and ILL is that anvils made of cBN can hardly be taken beyond 10 GPa without failure or damage after unloading. As for TC anvils, the pressure limit of single-toroidal anvils seems to be approximately 13 GPa, for double toroidal anvils approximately 16 GPa, depending on the type of TC (information obtained from various experiments at PEARL by the Edinburgh/Paris groups).

The pressure load characteristics of Figure 10.7 should be regarded as a guideline, i.e. valid for "standard" samples and loadings where the overall compressibilities (sample plus pressure transmitting medium) are similar. It is obvious that for samples with high compressibility and/or samples which show strong pressure-induces first-order transitions the pressure efficiency will be smaller. The same is expected if the filling fraction is not ideal (incomplete filling of pressure chamber) or if the ratio between the amount of sample and pressure transmitting medium is low.

There is a strong hysteresis between up and downstroke which makes measurements in downstroke risky and not recommendable. Similar to diamond anvil cell techniques, a recompression of a used gasket leads to a very low efficiency and an appreciable pressure increase is only observed at a load which is higher than the highest load attained in the previous upstroke run.

The choice of the anvil material depends entirely on the type of measurement, i.e. the scattering geometry and if contaminant Bragg reflections from the anvils can be tolerated. For "transverse" scattering (in through the anvil, out through the gasket) the choice is between TC and sintered diamond whereas for "longitudinal" scattering (in through the gasket - out through the gasket) cBN or sintered diamond is preferable, in particular for diffraction since TC produces a large number of contaminant Bragg reflections. For inelastic scattering on single crystals (such as done on triple axis spectrometers) TC is acceptable since such measurement focus on isolated points in (ω-q) space where Bragg reflections of TC rarely hamper the data collection.

10.3.2   Double-toroidal anvils

Here the sample volume is reduced by more than a factor 2 to approximately 30 mm^3 and the overall diameter of the gasket assembly is slightly smaller. The reduced sample volume allows pressures of 25 GPa to be reached rou-

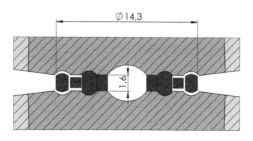

FIGURE 10.8
Gasket assembly in double-toroidal anvils using washer-type (upper) and encapsulating (lower) inner gaskets. Dimensions of the anvil profile are given in Appendix D.

tinely applying forces of almost 200 tonnes onto the anvils. These pressures can only be supported by sintered diamond (chapter 4.3). The higher neutron transparency of this material (chapter 4.3.5) compared to TC partially compensates the reduced intensity due to the smaller sample volume. As a result of this, the signal level observed in the transverse mode is very similar for the two types of anvils. The method of choosing gaskets having decreasing thickness with increasing diameter is also applied in double-toroidal gaskets: As shown in Figure 10.8, the thicknesses are such that the initial contact is only on the inner set of gaskets up to the first toroid.

The pressure-load curve for typical samples (as defined in the previous section) is illustrated in Figure 10.9. These data are based on three types of measurements: (a) a large number of measurements on ice (phases VII and VIII for which an accurate equation of state exists) from which a typical run was chosen. Encapsulating TiZr gaskets were used which ensure that the liquid water sample is properly sealed from the begin on. (B) a run where MgO was used as pressure marker in the same type of gaskets and 4:1 ME pressure transmitting fluid. (C) Samples of B_2O_3, SiO_2 and GeO_2 glasses in form of compacted pellets and washer-type TiZr gaskets. The pressure in these runs was determined from the shift of the Bragg reflections from the diamond anvils, see section 10.3.1, which was calibrated against pressure in a run with MgO.

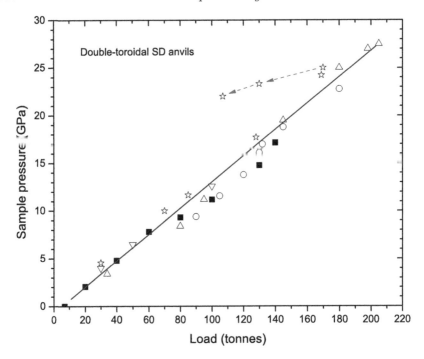

FIGURE 10.9

Typical pressure-load relation for double-toroidal sintered diamond anvil profile as shown in Figure 10.8. Open symbols correspond to runs with ice VII
samples, collected by the Paris/Edinburgh groups in washer-type inner gaskets (data provided by C.L. Bull). (■) corresponds to measurements on a
perovskite sample with MgO as pressure marker, using encapsulating inner
gaskets and ME as pressure transmitting fluid (data provided by M. Tucker,
ISIS). All data measured in upstroke (increasing load), except where indicated
by arrows.

As seen in Figure 10.9 the threshold for pressure increase is below 5 tonnes
since the initial contact surface is smaller, and this is typically the load initially applied to seal the gasket. The $P(F)$ curve is then approximately linear
till 10 GPa/100 tonnes and shows in most cases a kink in the slope at ≈ 100
tonnes, beyond which the pressure efficiency increases. As stated in the previous section, such a "calibration-curve" is a mere guideline and holds only
for loadings which have overall volume changes similar the samples included
in this plot. Nevertheless the data of Figure 10.9 indicate that the scattering
between samples and encapsulating as well as washer-like gaskets is only in
the order of $\pm 10\%$ in pressure. Clearly, the filling factor is the most deci-
sive parameter, i.e. a "bad" loading will deviate substantially from the curve
in Figure 10.9. Also, the P(F) curve will deviate from a straight line when-

ever there is a first-order transition to a significantly denser phase and the ultimately achievable pressure will be lower.

For "ordinary" samples, using the gasket-anvil setup shown in Figure 10.8 and TiZr gaskets, the pressure increases up to typically 25-30 GPa where the experiment is terminated by a violent blow-out, sometimes a few minutes after the load increase. Such events manifest themselves by a loud bang, a drop of the load by 10-30%, gasket material which self-ignites and which is violently ejected. Shrapnells of TiZr were observed to pass 0.5 mm of aluminium. It is needless to explain that severe safety precautions have to be taken in such experiments, i.e. proper shielding and protection of the immediate environment as well as strictly limited access to the pressure cell. Blow-outs at high load usually produce damages on the profiled anvil surface; for loads beyond 150 tonnes the chances to lose at least one anvil is probably around 50%. Experiments with CuBe as gasket material show anvil failures at similar pressures from which we concluded that the TiZr is not the determining factor for blow-outs. It is most likely due to a general weakness of the gasket-anvil geometry itself.

One might wonder about the pressure performance of double toroidal anvils made of other materials than sintered diamond. Tests carried out by the author indicate that the maximum pressure for TC is between 14 and 16 GPa, depending on the type of TC, and 11 GPa for cBN. From this it can safely be concluded that the only possible way to achieve pressures beyond 20 GPa is by using the anvils made of sintered diamond.

10.4   General remarks and observations

1. Approximate pressure determination

It is an inconvenience of opaque anvils that the pressure cannot be determined easily, in particular for disordered systems (glasses and liquids) which produce no Bragg reflections, or if the instrument cannot measure lattice constants accurately as it is often the case in inelastic neutron scattering. Nevertheless there is an easy way to determine approximately the maximum achieved pressure after unloading. From this, intermediate pressures might be determined from the forces on the anvils and appropriate interpolation schemes. The only hypothesis in this method is that one assumes that the pressure distribution across the gasket and sample is parabolic, see Figure 10.10:

$$P(r) = P_{max}[1 - \frac{r^2}{R^2}] \qquad (10.1)$$

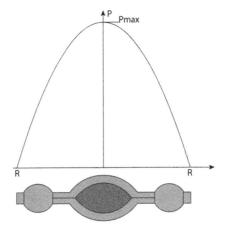

FIGURE 10.10
Simple model of pressure distribution across the sample-gasket assembly in toroidal anvils.

where P_{max} is the maximal pressure at the center of the sample chamber and R the radius of the gasket where the pressure is necessarily zero. Since the total force F onto the anvil is the surface below the parabola, integration of equation 10.1 between 0 and R gives:

$$P_{max} = \frac{2F}{A} \qquad (10.2)$$

where $A = \pi R^2$ is the cross section of the gasket. In principle the two quantities F and R can be measured, even during the experiment, and the pressure be determined by equation 10.2. Post-experiment measurements show that for "standard" samples equation 10.2 indeed gives pressure values about $\pm$ 10-20% close to the pressure measured by in situ methods. Standard samples are those for which the parabolic approximation 10.1 holds, i.e. those with compressibilities sufficiently high to give a continuous pressure variation at the sample-gasket interface.

2. Principle of massive support

There is the obvious question what could be the strategy to increase the achievable pressure *without* reducing substantially the sample volume and *without* increasing substantially the required forces which would require more massive load-frames and which would compromise the flexibility of the current PE devices. Note that low-temperature experiments rely on compact pressure cells which hardly can have a mass beyond 50 kg. Here are a few remarks and observations made over the last two decades by the Paris and Edinburgh groups on a large number of various anvil types similar to those shown in Figures 10.6 and 10.8.

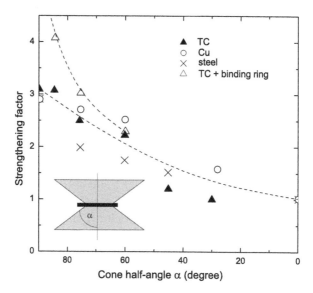

FIGURE 10.11
Strengthening factor in Bridgman-type anvils as a function of cone semi-angle α (inset), for different materials. TC = tungsten carbide. Adapted from Ref. [3].

Let us start with an important result valid for all types of Bridgman anvils, related to the *principle of massive support* [5, 95]. It states that a semi-infinite plane of material has a compressive strength considerably higher than the compressive strength obtained from finite-size test specimens, such as indicated in Table 4.2, for example. Compressive strength is determined by the onset of crushing of cylindrical or cubic test specimens. If instead a test is carried out on a large plate with a force applied on a relatively small area, the compression strength defined as the onset of indentation is considerably higher, basically because the material around the contact surface supports the material under compression. The gain in compression strength is called the strengthening factor, and it has been carefully investigated as a function of the cone angle in Bridgman-type anvils, see Figure 10.11. For a truncated cone this factor is reduced and by definition falls to unity when the cone angle is equal to zero. The strengthening factor has been determined experimentally on a number of materials (Figure 10.11) but has also been calculated [3]. It is clear from these investigations that large cone angles are favorable, and there is clear evidence for improvement if the anvil is supported by a steel binding ring. For a half-angle of approximately 80° as used in anvils shown in Figure 10.8 the strengthening factor is ≈ 3 and with a compressive strength of typically

6 GPa for TC as indicated in Table 4.2, the ultimate pressure capability for TC is expected to be 18 GPa. The same reasoning on the basis of sintered diamond with a compressive strength of 8 GPa (Table 4.4) leads to 24 GPa. This is in reasonable agreement with the maximal pressures where blow-out occurs using the anvil geometry described above and would indicate that the limiting factor is the anvil material rather than the gasket design. Bull et al. developed a modified anvil profile with an outer cone half-angle of 70° to give improved angular access for single-crystal diffraction [91]. As expected, the maximal achievable pressure for such a geometry decreases, to 7 GPa using TC as anvil material.

3. Pressure efficiency

The efficiency, i.e. the slope of the $P(F)$ curve, is governed by a delicate balance between the capability of the gasket to extrude and its capability to resist the pressure build-up in the sample: If the gasket cannot extrude (because the anvil profile prevents to do it), the efficiency will be low since the compressibility of the metal is small. On the other hand, if the extrusion is large (i.e. the gasket can flow outward easily), the pressure build-up will be small as well, and again the efficiency will be small. It is a recognized fact, based on tests on a large number of different anvil geometries, that the anvil profile of Figure 10.6 has high efficiency. For a given geometry of the pressure chamber (cup) and gasket material, the main parameters which determine the efficiency are (A) the overall diameter of the gasket, and (B) the height of the outer edge of the toroid groove (at a diameter of approximately 15 mm). The edge determines in a critical way the possibility for the toroidal gasket to extrude: if it is increased (by making the cone half-angle 90°, for example), the efficiency is strongly reduced.

In general, it should be pointed out that the efficiency of toroidal profiled anvils is impressive compared to diamond anvil cell (DAC) technology. To achieve 10 GPa on a 80 mm^3 sample, toroidal anvils need a force of 100 tonnes (Figure 10.9). For a DAC to achieve this pressure, the forces are 3 orders of magnitude lower (typically 0.1 tonnes), but the compressed sample volume is 5 orders of magnitude smaller (typically 10^{-4}-10^{-3} mm^3). Whereas efficiency is not critical for DAC applications, it definitely is for high pressure neutron scattering since the required forces determine the mass of the load frame and hence the science which can be addressed. Note that in the average, more than half of all neutron experiments involve low-temperatures for which the size of the pressure cell is critical.

11

Paris-Edinburgh cells II: Low and high temperatures

CONTENTS

11.1   General

The importance of studying high pressure phenomena at variable temperatures is obvious given that neutron scattering deals with issues such as magnetism, hydrogen ordering, and lattice dynamics which all depend strongly on temperature. The PE presses have the capability to cover the temperature range of 3.5 K to approximately 1000 K. The technical solutions are however not unique and depend strongly on the temperature range required, the kind of measurements and the scattering geometry.

11.2   Low temperatures

The key difficulty is to be able to operate the hydraulic ram even at low temperatures, which prevents the use of standard hydraulic oils which solidify below 240 K. The simplest solution for moderate low temperatures is the use of a 5:1 iso-n-pentane mixture which remains liquid to 110 K, even under pressure of a kilobar, see Appendix E. If simultaneously the O-ring seal (Figure 10.4) is replaced by a steel enforced PTFE seal, any PE press can operate down to 110 K, with the possibility to change the pressure at low temperatures.

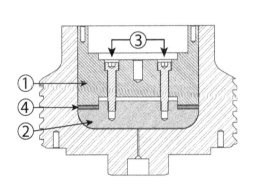 

FIGURE 11.1
Low temperature ram with Bridgman unsupported area seal [96]. (1) piston - upper part; (2) piston - lower part; (3) pre-compression bolts; (4) PTFE-Pb-PTFE sandwich seal.

For temperatures below 110 K the only possible hydraulic fluid is compressed helium which makes the sealing considerably more difficult. The so far best - but maybe not optimal - solution consists in a piston with an unsupported Bridgman seal, see Figure 11.1 [96]. The seal is pre-compressed by a set of bolts which ensures leak tightness at low load. The sealing package consists of a set of PTFE-Pb-PTFE rings (0.5-2.0-0.5 mm). Packages with indium instead of lead are another option but the extrusion of indium can be considerable at high load and indium tends to stick to the cylinder walls. Equipped with such a low temperature piston, several VX5 cells have been in use for the last 6 years at the ILL and PSI. Strain-gauge and in-situ diffraction measurements indicate that the piston is fully moveable down to at least 20 K, limited only by the freezing of helium. These measurement also indicate that whilst the friction produced by the Bridgman seal is negligible at 300 K, it is not at low temperatures below typically 180 K (Figure 11.2). This has a consequence for the variation of pressure during warming: If a certain load is applied at low temperatures and maintained, the effective load onto the anvils will increase during warming since the friction decreases progressively as the temperature increases. To avoid this effect the He-gas pressure needs to be reduced at low temperatures to the value indicated in Figure 10.7 for the pressure aimed for.

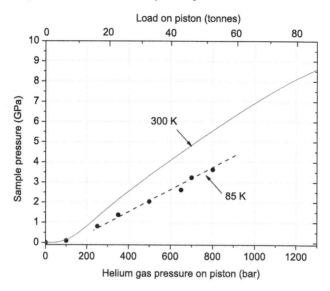

FIGURE 11.2
Pressure-load relation for VX5 ram with Bridgman unsupported area seal (Figure 11.1) and standard single toroidal anvils. Lines "300 K" and "80 K" correspond to compression at these temperatures. The 300 K line is identical to the one shown in Figure 10.7, the 80 K line is taken from Ref. [96].

11.2.1 Liquid nitrogen cryostats

The PE-cells needs specific cryogenic equipment since their masses and dimensions are incompatible with standard cryostats used for neutron scattering. There are two possible design strategies:

(A) Cooling the whole cell to the temperature of the sample. The advantage is that temperatures close to 77 K can be reached and controlled in a straight forward way, in the easiest case by simply suspending the cell above a liquid nitrogen bath. The inconvenience is that helium has to be used as hydraulic fluid which is more difficult to seal than more conventional hydraulic fluids.

(B) Cooling only the immediate sample environment (sample, gasket, anvils, backing plates) and keeping the load frame as warm as possible. The advantage is that the system is very reactive since the mass to be cooled is only a fraction of the total cell mass. Another advantage is that the hydraulic parts (piston seal) are sufficiently warm to avoid sealing problems. The inconvenience is that the heat leak into the anvils is sufficiently large to limit the minimum achievable temperatures to well above 77 K, typically to above 100 K. Also, temperature control is relatively complex since there is a continuous liquid nitrogen flow through parts of the cell.

Both methods have been applied on various high pressure beamlines and

 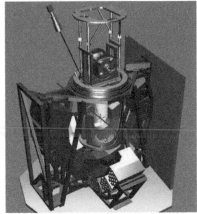

FIGURE 11.3
Liquid nitrogen cryostat for PE-press at PEARL (ISIS). Right: overall view showing pressure cell on cradle lifted out of the cryostat (schematic). Left: picture of cradle carrying the PE press. Source: W.G. Marshall & M.G. Tucker, ISIS.

are still under development. Figure 11.3 shows the setup at PEARL (ISIS) used since 1998 which adopted method (A). It arouse out of a simple cryo-container operating before at the POLARIS station [97]. Here the cryostat is a vacuum isolated stainless steel-aluminium container of 400 mm inner diameter and 565 mm depth covered with a perspex lit. The cell is placed with its thrust axis horizontally on a cradle which has a mechanism to rotate the cell. The cell on its cradle is then lifted into the cryostat and cooled by spraying liquid nitrogen on it. This method allows the 50 kg V3 cell to reach 150 K in less than an hour, and 80 K in less than 2 h. The temperature is then maintained by the cold gas from liquid nitrogen which has accumulated on the bottom of the cryostat, and controlled by heating using two 150 W band heaters attached around the cell body. This method allows to reach temperatures close to 77 K, in practice 80-85 K for routine data collection.

Although this cryogenic setup might appear somewhat unsophisticated it has one uncontestable strength compared to all other setups discussed further below: The cell is in a vessel which is at atmospheric pressure and easily accessible even at low temperatures. This allows to recover samples at low temperatures and extract them from the cell. Such manoeuvres have been essential in the study of ice for example, and can at present only be carried out with the cryogenic setup at PEARL. A weakness of the system are frequent leaks of the vacuum insulation which is eventually related to the large overall size of the cyrostat. Also, the rotation mechanism almost doubles the mass to cool and warm, increases the thermal inertia, and makes the handling of the cell more cumbersome. A similar, but more compact system with a more sophisticated cooling system would be beneficial for most of the routine experiments.

Figure 11.4 shows a cryogenic system operating since 2008 at the PEARL station. The conceptual difference is that it is not a bath-cryostat but cooling is achieved by circulating liquid nitrogen through cooling rings which are in thermal contact with the anvils. In this way only the immediate environment of the sample is cooled, i.e. the anvils, and to a lesser extent also the backing seats. The thermal heat leaks are minimized by using backing seats and plates made of zirconia, a ceramic with high compression strength and low thermal conductivity. As a result of this, the system is very reactive, cooling to the base temperature (currently 110 K) is possible within 45 minutes. Heating is achieved by resistive heaters incorporated in the anvil steel binding rings. These 240 W heaters allow at the same time heating to 500 K as will be discussed further below.

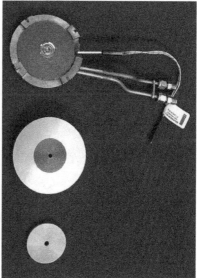

FIGURE 11.4
Variable temperature (v-T) set-up at the PEARL beamline (ISIS). Left: over-all view with the V4 PE cell suspended at the vacuum flange. Right: Details of anvils with attached cooling rings and heater leads (upper), zirconia backing seats (middle) and backing disk (lower). Picture courtesy of W.G. Marshall, ISIS.

The basic technical difficulty of this system is to control the temperature by a coordinated control of liquid nitrogen flow and resistive heating. Since the hydraulic ram is maintained at temperatures above 110 K, an iso-n-pentane mixture (Appendix E) can be used as hydraulic fluid which avoids the recurrent sealing problems with helium. The second major difference with respect to the bath-cryostat concept is that the cell is suspended on a flange and in vacuum which serves as a thermal insulation. Obviously, this concept makes the design of such a cryostat extremely simple and compact. But, as mentioned above, the sample is inaccessible at low temperatures and cannot be recovered under these conditions.

FIGURE 11.5
Nitrogen cooling system at the SNAP diffractometer (SNS, Oak Ridge). Left: fully assembled; right: close-up view on the VX5 pressure cell. Picture courtesy of J. Molaison, ORNL.

The cryogenic system available at SNAP (SNS) for VX5 type PE cells follows a very similar concept. Here the cell is also suspended in vaccuum on a top flange and the anvils cooled by circulating liquid nitrogen. The heat input from the warmer cell is minimized by adding thin Garolite G10/FR4 discs between the backing plates and the piston and breech spacers. The present design (Figure 11.5) can cool a sample to 85 K in 30 min and uses a Lakeshore-relay-controlled cryogenic solenoid valve to maintain temperature within ± 5 K. The preferred scattering geometry at SNAP is "longitudinal" (see Figure 10.5) with the thrust axis of the cell vertical, as shown in Figure 11.5. This orientation is more favorable for the design of cryostats which tend to be cylindrical, which means that the cryogenic system can be built in a more compact way. This system will be equipped with a high-capacity closed-cycle refrigerator linked to the anvils which will allow to cool samples to below 80 K.

11.2.2 Helium cryostats

The first cryostat for PE cells operating to liquid helium temperatures and below was presented in 2006, see Figure 11.6 [98]. An identical system is operating at the PSI since 2007. The experimental possibilities to carry out high pressure neutron scattering in the 10 GPa range and down to 3.5 K enables obviously entirely new physics, in the field of magnetism for example. This system is designed around a 35 kg VX5 cell cooled by a closed-cycle refrigerator with the option to pre-cool rapidly by immersing the cell in liquid nitrogen. To reach such low temperatures the entire cell has to be cold. Contrary to the PEARL v-T system and the SNAP design, the cell is not in vacuum but in a 10-100 mbar He exchange gas which provides the thermal contact to the inner shield and insures minimal temperature gradients. This choice has the important advantage that small He-leaks through the Bridgman seal (which occur frequently) do not immediately destroy the thermal insulation of the cryostat and thereby produce a massive heat leak. Cooling the cell to base temperature using only the CCR would require approximately 30 hours. In practice, the cell is pre-cooled by flooding the inner chamber with liquid nitrogen which allows to reach 77 K within less than 1 hour. The nitrogen is then evacuated and the remaining cooling done by the CCR, a procedure which takes approximately 6 hours. The operation of the pneumatic (He-driven) ram is only limited by the freezing of helium which can be at temperatures of up to 20 K at the highest gas pressures (see Appendix E). Even if the helium in the ram and tubing is solid, the cell can be further cooled without danger. Heating across the melting line is however more delicate and potentially dangerous for the equipment since it causes a sudden pressure increase if the expanding fluid helium cannot be evacuated quickly enough through the gas handling system.

11.3 High temperatures

High pressure–high temperature (high P/T) techniques can be divided in two categories with quite distinct technical aspects: (A) external heating where the sample plus the environment, including the anvils and in the extreme case the entire cell is heated. (B) internal heating where only the sample is heated and the anvils (as well as the rest of the cell) are essentially cold. As will be shown in the following paragraphs, the choice between one or the other method is dictated by the required temperature: Up to 500 K external heating is a technically simple solution whereas for higher temperatures the only option is the technically more challenging internal heating.

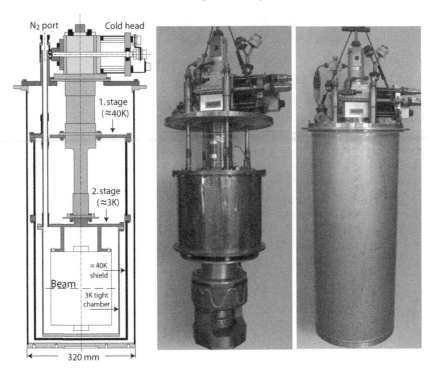

FIGURE 11.6
Cryostat at ILL to cool a VX5 PE cell to 3.5 K. Cross section (left), without
vacuum and thermal shields (middle), and fully assembled (right). Courtesy
of ILL, Grenoble.

11.3.1   External heating

The ultimate limit for the steel used in all PE cells being approximately $300°C$,
external heating could be achieved by simply heating the entire cell (which
was accidentally done). However various elements cannot be heated to such
temperatures, among others the O-ring seals, the hydraulic oil, and the cadmium shielding on the anvils. The strategy in at least two current high P/T
PE setups has therefore been to heat only the anvils and improve the thermal
insulation with the rest of the cell.

   Figure 11.7 shows the setup used at PSI and ILL for angular dispersive
diffraction [99]. Here only the anvil die is heated through a commercially available 80 W cartridge heater placed in a bore at the back of each anvil. The
dies are thermally insulated in radial direction by a zirconia sleeve, and axi-
ally by backing seats with zirconia dies. Zirconia has a compression strength
of ≈ 2.2 GPa and a thermal conductivity of 2-3 W/mK, compared to TC

with 5 GPa (Table 4.2) and 30 W/mK (Table 4.1), respectively. The zirconia seat can hence support the required loads of 100 tonnes (Figure 10.7) if the contact surface with the anvil is larger than 4 cm². There are cooling rings radially attached to the anvils and backing seats where air is circulated during experiments, and water to cool the anvils rapidly. This assembly allows to reach 450 K in 20 minutes when the cell is in air, and 550 K when the whole cell is in vacuum [99]. The temperature is measured so far by two thermocouples attached to each die, at approximately 2 mm from the edge of the gasket. Calculations show that due to the high thermal conductivity of cBN (70 W/mK, Table 4.3), the temperature gradient across the cBN die is at most 5 K at 500 K. With continuous operation at 400-500 K and air-cooling of the anvils the cell reaches a temperature of approximately 80°C after 10 hours, with the anvils remaining below 130°C, which is often considered the upper limit for cadmium shielding. Operation over weeks of continuous data collection at PSI and ILL proved the reliability of such a system. The same setup can be used with sintered diamond anvils. The pressure performance seems to be only limited by the decreasing strength of TiZr at high temperatures. Figure 2.4 indicates that the tensile strength of TiZr decreases by approximately 15% between 300 K and 500 K. Given the fact that there are no alternatives zero-scattering gasket materials, external heating in angular dispersive neutron scattering is intrinsically limited by the TiZr.

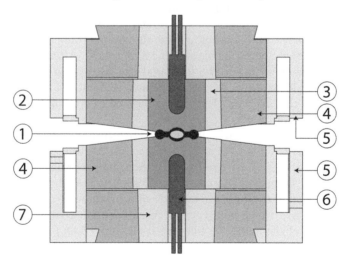

FIGURE 11.7
Cross section of gasket-anvil assembly for high P/T measurements using external heating. (1) TiZr gasket; (2) anvils (cBN); (3) thermal insulation sleeve (zirconia); (4) steel binding ring; (5) cooling rings; (6) cartridge heaters, (7) backing seat with zirconia die and steel supporting ring. Adapted from Ref. [99].

PEARL's v-T setup (Figure 11.4) applies a similar strategy for external heating, i.e. heating only the tungsten carbide anvils with thermal insulation achieved by zirconia backing seats. However, the heating elements cannot be located in the dies (since diffraction on PEARL is done in the transverse mode) but incorporated in the steel binding ring. The advantage of this scattering geometry is that metals other than TiZr can be used for gaskets with properties more adapted to high temperatures. The inconvenience of this type of heating is that the radial support of the steel binding ring decreases with increasing temperature due to the larger thermal expansion of steel compared to tungsten carbide, or any other anvil material. Also, the cadmium shielding on the steel binding ring has to be replaced by gadolinium which supports considerably higher temperatures.

11.3.2   Internal heating

The development of internal heating was guided by considerable prior experience from high P/T material synthesis and x-ray synchrotron diffraction measurement where profiled Bridgman anvils have been used since a long time (Refs. [90, 85] and refs. therein). Whatever the details, all designs use resistive heating by a graphite tube inside the pressure chamber to reach temperatures up to 2000 K, in routine practice rather 1300 K. Since the anvils need to remain essentially at room temperature, the furnace has to be surrounded by thermally insulating material, i.e. some kind of ceramics. This insulation needs typically 50% of the available volume, i.e. the sample volume has to be reduced if the same pressure performance shall be maintained. Alternatively, the overall gasket dimensions have to be increased which causes a dramatic decrease of the achievable pressure. A major technical difficulty of internal heating is also related to the fact that the gaskets must be electrically insulating which excludes the use of metals.

The historical first attempt in high P/T neutron scattering using internal heating is reported by Zhao et al. (Los Alamos) [102], where simultaneous P/T values of 7 GPa/1000 K are cited and a diffraction pattern at 4.2 GPa/1025 K is shown. This setup used cupped toroidal anvil (a scaled-up version of those shown in Figure 10.6) and gaskets made of a machinable ceramic, zirconium phosphate (ZPM), which has very low thermal conductivity (1 W/mK). The spherical cups of the anvils are filled with zirconia (ZrO_2) and electrical contacts to the anvils are provided by thin platinum wires passing through the ZrO_2 caps. The thickness of the ZPM gasket was chosen to be surprisingly large, probably with the aim to increase the sample volume. As a consequence, the extrusion of the friable material is considerable, in particular at the start of the compression, and only partially prevented by a Teflon ring around the ZPM gasket. In situ temperature measurement were attempted with the use of a thermocouple through the gasket but are reported to be challenging, again because of the extreme deformation of the gasket. This coincides with experience obtained on similar setups for x-ray synchrotron diffraction (Y.

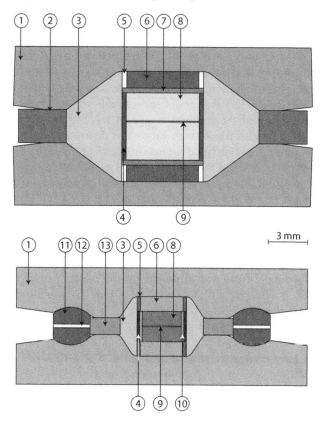

FIGURE 11.8
Gasket assembly for high P/T diffraction using internal heating and conoidal anvils. Anvils. (1) anvils; (2) Teflon ring; (3) pyrophyllite gasket; (4) graphite furnace; (5) electrical contacts (stainless steel); (6) MgO disc; (7) graphite disc; (8) sample; (9) Ta or Hf foil; (10) MgO sleeve; (11) CuBe gasket; (12) MgO glue; (13) pyrophyllite gasket. Adapted from Refs. [100] and [101].

Le Godec, priv. communication) where in most of the routine runs the temperature measurements rely on the measured injected power and a suitable calibration curve obtained in an identical setup.

Figure 11.8 (upper) shows a setup developed by the Cambridge group [100, 103] which arouse out of a high P/T setup used previously for x-ray diffraction (I. Makarenko, Univ. P&M Curie, Paris, internal report). This setup uses Bridgman anvils with conoidal cups ('conoidal anvils') and a pyrophyllite gasket, radially supported by a Teflon ring. The conoidal shape of the pressure chamber with a flat bottom is better adapted to the geometry of the furnace and simplifies the electrical contact to the anvils using two stainless steel rings. The sample volume in this cell is approximately 50 mm^3 at the start of the experiment and hence not considerably smaller than in the 300 K setup. With a graphite furnace of 0.2 mm thickness the resistivity measured at the power supply is typically 0.01-0.05 Ohm, i.e. a current of typically 100 A is needed to inject the required 100-200 W to reach temperature in excess of 1000 K. For accumulation times of several hours water cooling rings around the anvils keep the cell at ambient temperature. Similar to the Los Alamos design, the gasket thickness is very large and decreases rapidly in the initial compression phase without building up substantial pressure. Again the extrusion of the gasket material is considerable. The particular anvil profile (lack of a toroid) leads at high pressure to a situation where the gap between the anvils is almost closed (less than 0.5 mm) and hence the diffracted intensity strongly reduced.

The setup of Figure 11.8 (upper) is since 2003 a user facility at PEARL (ISIS) where routine high P/T measurement to simultaneously ≈6 GPa/1000 K can be carried out. An important development related to this high P/T cell is the in situ temperature measurement by the nuclear absorption resonance method using a thin (50 μm) Ta or Hf foil (item 9), a method which will be described in section 11.3.3.

A general drawback of both the Los Alamos and the Cambridge designs is that the gasket material (ZPM or pyrophyllite) is very friable and cannot provide any lateral support against extrusion. These materials lack a critical property required for any gasket material: ductility, i.e. the ability to flow without fracturing. Therefore, the gasket diameter increases rapidly with load as its thickness decreases substantially, entailing a massive loss in intensity as the anvils close and the diffracted beam has to cross more and more gasket material. The design shown in Figure 11.8 (lower) removes this inconvenience [101]. The strategy in this design is (A) maintain a tordoidal anvil profile since it ensures a relatively large gap between the anvils (approximately 0.5 mm at the highest load), and (B) use a conoidal pressure chamber since it is adapted to the geometry of the furnace and the associated elements. A novel aspect in this design is the use of a metallic toroidal gasket, consisting of two preformed parts separated by a 0.2 mm layer of MgO glue. Such a gasket can be made easily and remains electrically insulating even under a load of more than 100 tonnes. The extrusion is perfectly controllable and reproducible which means that the signal-to-background ratio does not vary so dramatically than in the previously described designs. Although the sample volume is smaller than in the Los Alamos and Cambridge designs, the recorded intensity is approximately equal, if not larger because of the reduced absorption. The setup has been used in measurements on iron where clean diffraction patterns at the α-γ-ε-triple point at 8.2 GPa/650 K were obtained [101]. Diffraction patterns of similar quality were recorded up to 9.1 GPa at 300 K and 1000 K at 7 GPa.

11.3.3 Temperature measurement by Neutron Resonance Spectroscopy (NRS)

This is an elegant method to measure temperatures of samples under pressure at distance, without the use of thermocouples, avoiding therefore all the aforementioned difficulties. It reduces temperature determination to a simple absorption measurements which take a few minutes and gives an absolute precision of typically ± 20 K. It should be regarded as a major progress in high P/T condensed matter research, and high P/T neutron scattering in general. The method was first proposed by Fowler and Taylor [104].

This method exploits the fact that many elements have sharp neutron absorption lines ("resonances") in the 1-10 eV range, hence at relatively short wavelengths of 0.001-0.1 Å. This part of the energy/wavelength spectrum is not exploited in diffraction measurements and is at the same time accessible with standard detector technology. It can be shown that the width of the resonance line is enlarged by the Doppler effect and given by [105, 106]:

$$\Delta = \sqrt{\frac{4mME_Rk_bT}{(M+m)^2}} \tag{11.1}$$

where E_R is the resonance energy, m and M are the masses of the neutron and the target nucleus, respectively, and k_b the Boltzmann constant. Therefore, the temperature can, in principle, be derived from the measured width of the absorption line. Strictly speaking, formula 11.1 is only valid for a free gas or a classical solid with a Maxwellian velocity distribution of the target nuclei. However, it can be shown that equation 11.1 is even correct for bound nuclei in a crystal lattice provided that the temperatures is well above the Debye temperature T_D. A simple estimation shows that the pressure dependence of the line width (through the change in Debye temperature) is negligible, i.e. the Doppler-broadening depends only on temperature. Fortunately some refractory metals like Ta and Hf have quite low Debye temperatures (T_D=252 K for Ta and 240 K for Hf) and nevertheless high melting temperatures (3290 K for Ta and 2506 K for Hf), well beyond the capability of the high P/T cells in Figure 11.8. Other possible choices are [77]Ir, [75]Re and [103]Rh, all of them occurring in natural material. This flexibility gives enough possibilities to avoid interference with samples containing the same elements.

Figure 11.9 shows an example of a neutron transmission measurement through a high P/T setup as described in Figure 11.8. The two strongest absorption bands at 4.16 eV and 7.63 eV are caused by the tungsten carbide anvils, i.e. are due to [186]W and [183]W, respectively. To derive accurate temperatures from the Ta and Hf lines (located at 1.098 eV, 2.388 eV, and 10.34 eV) using equation 11.1, the measured data are fitted to analytical expressions for resonance lines (essentially the Breit-Wigner formula). The fit incorporates tabulated parameters of the lines (position and natural width) and takes into account the resolution of the instrument. With accumulation times of typically 20 minutes, the statistical error translates into a tempera-

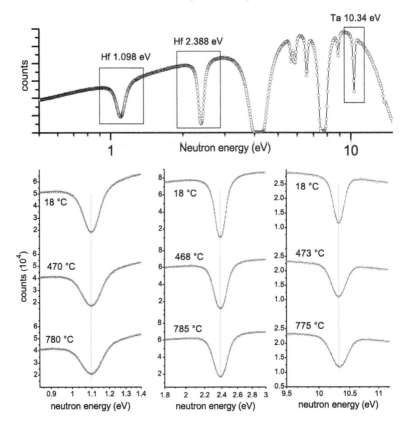

FIGURE 11.9

Neutron transmission through a gasket-anvil assembly similar to Figure 11.8 including 50 μm foils of natural Hf and Ta. Upper: Overall spectrum obtained at 300 K with absorption lines from [177]Hf 1.098 eV, [177]Hf 2.388 eV, and [181]Ta 10.34 eV. Lower: close-up view of the resonance lines at three different temperatures. Lines are fits to the data (circles). Adapted from Ref. [106].

ture uncertainty of less than ± 20 K for a single resonance line. Since there are at least two lines available, the temperature uncertainty could even be reduced. The temperature difference indicated by different resonance lines is usually small and can be taken as the real (systematic) uncertainty. Practical experience with this method and using Ta and Hf as temperature sensor shows that the requirement of $T \gg T_D$ is not so stringent as one might suspect, i.e. even temperatures below ambient are surprisingly accurately determined by NRS.

High pressure NRS measurements can be carried out with V4 or VX4 (Figure 10.4) load frames which allows transmission measurements. The neutron beam has therefore only to cross the elements shown in Figure 11.8, i.e. two anvils, MgO and graphite discs, the sample and the Ta/Hf foils. Accumulation of absorption and diffraction data are usually carried out in parallel.

11.3.4 Internal heating: General remarks and conclusions

Given the complexity and the geometry of the cell assemblies shown in Figure 11.8 it is obvious that a transverse scattering geometry (Figure 10.5) with a diffraction angle 2θ close to $90°$ is perfectly adapted to give clean diffraction patterns, and practical experience demonstrates this very well. A longitudinal scattering geometry would unavoidably produce an enormous amount of contaminant scattering of the various elements in the equatorial plane of the anvils, even with the use of more sophisticated radial collimators. The availability of zero-scattering materials with the required thermal and electrical properties is rather unlikely.

It is therefore clear that high P/T neutron scattering (in particular diffraction) is a domain where pulsed sources have a considerable advantage. The ease of measuring temperature under extreme P/T conditions by NRS (see section 11.3.3 above) is an indisputable strength of such sources. For these reasons it seems unlikely that continuous sources will ever be able to be competitive with pulsed facilities in the area of high P/T neutron scattering.

The techniques illustrated in Figure 11.8 have so far only been applied to pressures up to 9 GPa. There is no fundamental reason why this cannot be extended to pressure to 20 GPa given the ample experience already available from 300 K experiments. For this purpose the sample volume would have to be reduced to not more than 5 mm^3 to keep the forces below 200 tn. It is therefore clear that such measurements can only be carried out at high-flux spallation sources, and in fact constitute a major scientific case for the construction of such facilities.

12

Paris-Edinburgh cells III: Ancillary equipment

CONTENTS

12.1 Oil and gas compressors

Operation of PE presses require a hydraulic or pneumatic system which can generate a pressure of up to 2500 bars, in routine operation usually up to 1800 bar. Depending on the type of cell, the volume filled by the hydraulic fluid inside the cell is between 5 cm^3 (VX1) and 50 cm^3 (V7). To this has to be added the volume in the tubing up to the high pressure pump which is typically 10-50 cm^3.

The simplest solution consist of commercially available manual oil pumps that provide pressures up to 3000 bar, though their reliability beyond 2000 bar is questionable. Some of these pumps can be used with various fluids, including pentane.

Automatic syringe pumps are a more sophisticated solution. Their capacity in terms of compressed volume depends obviously on the maximal pressure, but for a pressure of 2000 bars (which covers all applications), pumps with V=60 cm^3 are commercially available for a cost of typically 25 kEuro. Such equipment ensures local or remote pressure control, including ramp-drives. Such equipment is very useful if the cell temperature fluctuates and hence the pressure in the hydraulic fluid changes due to the thermal expansion, i.e. in moderately-low and high temperature experiments. The inconvenience with all types of syringe pumps is the reload when the piston is at the end of the stroke, a situation which can frequently occur since the volume in the pump is approximately equal to the volume in the cell. There are pumps which allow to do that automatically but at considerable higher cost, complexity, and weight. A useful mode of operation is to equip a standard syringe pump with a small manual pump of a few 100 bars capacity which serves only to purge the hydraulic system and for initial compression to typically 100 bar. Further compression is then done automatically and experience shows that a single stroke of 50 cm^3 is sufficient for most of the experiments, i.e. there is no need for a reload-procedure. Syringe pumps can usually be used with a variety of liquids, including pentane, which allows operation of PE pressure cells to approximately 120 K.

As for low temperature operation, the cells require a pneumatic system using helium as hydraulic fluid. Gas pressure of up to 3000 bars are readily provided by membrane compressors, available from a few companies worldwide. It is recommendable to use a double-headed compressor to be able to work with low intake pressures. The main difficulty is automatic pressure control when pressure release is required, which needs the controlled opening of release valves at He pressures of typically 1000 bars. The ILL has designed recently such a system which allows remote control of He pressure and which is now routinely applied for controlling forces on PE presses [107].

12.2 Loading clamps

Loading clamps are devices which include the anvils and backing plates and which allow to apply and lock a force sufficient to seal the gasket, typically between 5 and 20 tonnes. Once loaded, the clamp is then inserted into a PE press for the experiment and further compression. The clamps are hence adapted to the PE load frames. Figure 12.1 shows two types of clamps adapted to the VX3/VX4 and V7 cells. A similar design for VX5 cells uses anvil dies with standard profiles but smaller backing plates, the VX5 clamp anvils with standard profiles but smaller steel binding rings and smaller backing seats.

In the simplest case the clamps might be loaded by hand tightening the locking ring, a procedure which normally does not seal the gasket but which simply allows to protect the sample from air or moisture during the transfer to the load frame. In general, however, the sample is loaded with a force of at least 3 tonnes onto the anvils using a rig and a small hydraulic ram, equipment which fits even in most glove boxes. Once the force is applied the locking ring is firmly tightened. Experience shows that at least 2 tonnes are necessary to ensure a vacuum tightness and 5 tonnes to ensure that a gas sample at 2 kbars does not escape.

Loading clamps were developed and regularly used with PE presses as early as 1999 [108]. Their importance cannot be underestimated, for various reasons. The use of such clamps allows to separate the procedure of sample loading - which can be time consuming, difficult, or might need safety precautions - from the data collection with its various restrictions on time and location. In this way a series of samples can be loaded into several clamps in advance, at a remote and adapted location, in a glove box in the home institute for example. This translates in a considerable gain in time and efficiency since the change

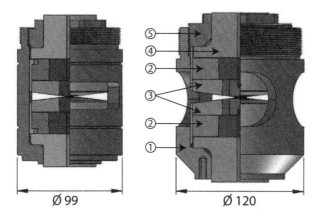

FIGURE 12.1
Loading clamps for VX3 (left, two columns), and V7 (right, four columns).
(1) clamp body; (2) backing plates; (3) anvils and gaskets; (4) steel seat; (5)
locking ring.

of sample, i.e. change of the clamp, is in most cases a question of minutes.
The use of loading clamps will become more and more frequent also because of
safety concerns and the ever increasing restrictions on sample handling close
to the neutron beam lines. Finally, the windows of the loading clamps can be
covered by an Al-sleeve which allows to confine the sample in case of blow-
outs and dispose the clamp, i.e. avoiding the risk of contaminating the entire
pressure cell and its environment.

It is predictable that the use of loading clamps will make high pressure
neutron scattering more and more similar to the mode of operation with diamond anvil cells, with users traveling with preloaded samples between various
neutron sources and using the local PE press as load frame.

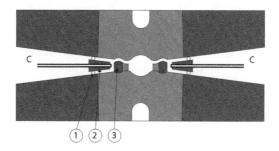

FIGURE 12.2
Condensing gas into toroidal anvils at low temperatures. (1) In-seal; (2) Al-
ring; (3) gasket; (C) capillary. Adapted from Ref. [109].

An example for applications of clamps is the loading of cryogenic gases as illustrated in Figure 12.2. In this case the clamp is cooled with the capillaries connected to a gas handling system which provides a gas pressure between 1 and 10 bars. Once the sample has condensed and entirely filled the pressure chamber a load of 5-10 tonnes is applied to the clamp using a hydraulic ram and locked by tightening the lock-ring. After warming back to room temperature the Al-ring is removed. This method has been applied for loading ammonia [110], methane [111], and oxygen [112] with success rates close to 100%.

12.3   High pressure gas loading

The use of gas loading in PE presses is motivated by essentially the same reasons as in diamond anvil techniques: (1) The possibility to use compressed gases as pressure transmitting media, and (2) the possibility to load gaseous samples which cannot be loaded cryogenically, for example gas mixtures. From what is explained in chapter 13.4.4 it is clear that a gas loader has to be able to load under pressures of at least 1500 bar, otherwise the limited volume change under compression will limit the achievable pressure. Note that the standard single-toroidal profile with its 1.5 mm cup depth can produce a volume reduction V/V_0 of at most 0.4. This situation is hence very different to diamond anvil cells where the sample volume can be reduced virtually to zero.

It is obvious that the gas cannot be loaded directly into the pressure cell given its dimensions, but instead into an intermediate device, i.e. a loading clamp as shown in Figure 12.1 or similar. Bocian et al. [113] presented a gas loader operating up to 1500 bars which works in conjunction with a loading clamp where the force is stored by a latch-mechanism. This device can be used to load any inert gas as demonstrated on single-crystal neutron diffraction measurement where argon was loaded as pressure medium [114].

However, loading gases at higher pressures would be very desirable since the volume contraction of most gases under compression is considerable. To illustrate this: for Ne, Ar, or N_2 the pressure required at 300 K to produce the density the liquid would have at 1 bar and low temperature is in all three cases above 2000 bar. And even if this can be achieved, subsequent compression still requires a volume contraction of typically 50% to reach 5-8 GPa, see Figure 13.2. It would also be highly desirable to load hydrogen/deuterium samples. Figure 12.3 shows a design which allows such loadings. The pressure vessel is designed for 3000 bar and was built around the loading clamps of Figure 12.1. The essential element of the locking mechanism is a shaft (3) which applies a force of up to 30 tonnes onto the anvils using a hydraulic ram (1). Once the force it applied (and the gasket sealed), the shaft is rotated and thereby

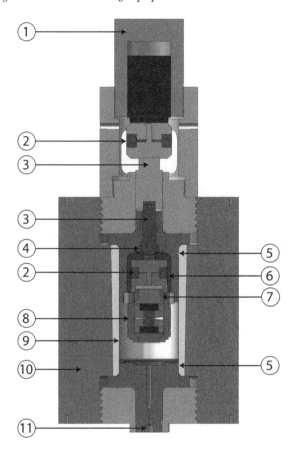

FIGURE 12.3
Pressure vessel to load gases under pressure to 0.3 GPa into PE loading clamps. (1) hydraulic ram (50 tn); (2) ball bearings; (3) push & rotation shaft; (4) dynamic seal; (5) static O-ring seals; (6) cap spanner; (7) locking ring; (8) loading clamp; (9) CuBe liner; (10) cylinder; (11) gas inlet. Spacers to reduce dead volume not shown. Source: S.K., unpublished.

tightens the locking ring (7) of the loading clamp (item 5 in Figure 12.1) via a cap spanner (6) attached to the shaft. Hydrogen compatibility of the vessel is achieved by using CuBe for the liner and the end plugs. The minimum dead volume of this device is approximately 100 cm^3 which is not negligible for expensive gases such as deuterium or deuterated gases in general. However, it is easy to recover most of the gas by releasing it back to a compressed gas bottle and ensuring that no contamination can occur during this procedure.

12.4   Anvil rotor

A recurring problem in high pressure single crystal studies is the shading imposed by the tie rods of the press which reduces the accessible Q-space and hence the data quality. Such difficulties occur for example in high pressure phonon dispersion measurements, and to a more serious extent in single crystal diffraction studies. A solution to this problem is a mechanism which allows co-rotation of the two anvils under load, a technique where there is already experience from x-ray micro-tomography studies using Drickamer-type anvils [115], and sheer deformation studies using profiled Bridgman anvils [116]. Figure 12.4 shows an anvil rotator for a V4 PE press developed for diffraction studies at D9 (ILL) and SXD (ISIS) [117]. A torque of 1.5 Nm provided by a stepper motor is translated by a 1:160 commercial gearbox to a final torque of 370 Nm onto the anvils. This mechanism allows a maximum speed of rotation of 1.35°/sec. A key feature are the custom-made oil bearings which produce a friction low enough to be handled by the rotation mechanism, even at loads of up to 100 tonnes. If necessary, the rotation mechanism can be taken off during the experiment to give full access to the 4 x 66° openings between the tie rods of the press.

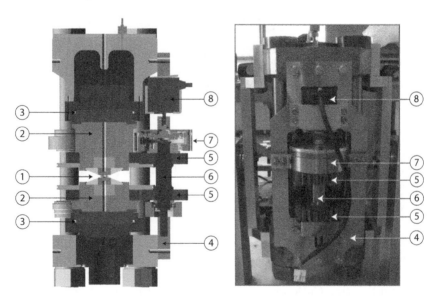

FIGURE 12.4
V4 PE cell with anvil rotator, after Fang et. al. [117]. (1) anvils; (2) anvil support; (3) hydraulic bearings; (4) side frame; (5) pinion; (6) side shaft; (7) gear box; (8) stepper motor. Adapted from Ref. [117].

13

Pressure determination and pressure transmitting media

CONTENTS

In contrast to light scattering experiments in diamond anvil cells, high pressure neutron scattering in the multi-GPa range uses in almost all cases opaque anvils which exclude fluorescence techniques for the pressure determination. This is often considered a major drawback of opaque anvils, but is in practice for most measurements a minor problem, for various reasons:

1) Approximate pressure values are known for the applied load and calibration curves as given in chapter 10. These are usually sufficient to guide an experiment. In many experiments the pressure coefficient of the physical property x under investigation (such as fractional atomic coordinates, phonon frequencies etc.) is such that the statistical error in x outweighs the uncertainty in pressure. In other words an increased precision in the pressure will not significantly increase the precision in the pressure coefficient of x. If a comparison with theory is sought, then the relevant parameter is anyway not pressure but volume (or volume change), which can in most cases be accurately determined.

2) A large part of neutron scattering studies deal with diffraction on crystalline samples for which the equation-of-state (i.e. the volume-pressure relationship) has already been determined by other means, for example x-ray

synchrotron diffraction or ultrasonic measurements under pressure. It should be noted that the determination of pressure by the sample itself (using its EoS) is the most accurate method since it avoids the problem of pressure gradients which might otherwise exist between the sample and the pressure marker. Even if the EoS is not known, approximate bulk moduli and derivatives can be estimated knowing the value for similar type of solids and empirical rules.

3) If the EoS is not known it is always possible to obtain it rapidly by a separate measurement with the pressure marker. But it should be said that it is not the purpose of high pressure neutron diffraction to establish accurate equation of state of materials.

The real difficulties appears hence only in studies of liquid and amorphous material where no Bragg scattering occurs and the only way to determine/estimate pressure is by the load and some calibration curve or a measurement with a pressure marker.

Requirements for neutron diffraction pressure markers are:

1. Inertness, if possible even to high temperatures, to avoid reaction with the sample or the pressure transmitting medium

2. Simple structure to avoid excessive overlap with Bragg reflections of the sample, maximize the scattered intensity, and allow to determine pressure with a few or a single reflection.

3. Low bulk modulus to have high sensitivity in the GPa range.

4. No phase transitions to at least 10 GPa.

5. Good neutron scatter

6. Low neutron absorption cross section

Cubic solids with only few strongly scattering elements are clearly favorable candidates.

13.1 Equations-of-state (EoS)

Experimentally determined compression data are usually fitted to some P(V) relation known as the *equation-of-state* (EoS)[1]. These contain — apart from the volume at ambient pressure V_0 — only two adjustable parameters:

The bulk modulus at ambient pressure B_0:

$$B_0 = -V_0(\frac{\partial p}{\partial V})_0 = -(\frac{\partial p}{\partial lnV})_0 \qquad (13.1)$$

and its pressure derivative at ambient pressure B_0':

---

[1]For an introduction and general review on this topic see: *Isothermal EOS functions for solids* by K. Syassen, MPI Stuttgart (unpublished).

$$B_0' = (\frac{\partial B}{\partial P})_0 \tag{13.2}$$

Note that B_0' is dimensionless. Experience shows that $B_0' \approx 4$, in very rare cases it is smaller than 2 or larger than 8., see section 13.3.

Equations-of-state are known today for thousands of elements and compounds, determined mainly from synchrotron x-ray diffraction and ultrasonic measurements. With these data in hand, the pressure of a sample can be directly determined from the measured lattice parameters, without the use of pressure markers. In principle this is the most accurate pressure reading since it gauges the pressure the sample is subjected to (and not that of the pressure marker which might be different due to pressure inhomogeneities), but the method relies obviously on the accuracy of the available EoS. With the ever increasing bulk of EoS data, pressure determination in neutron scattering using opaque anvils has become a minor problem.

The most widely used equations-of-state are:

1. The Murnaghan equation-of-state (M-EoS):

$$P = \frac{B_0}{B_0'}[(\frac{V_0}{V})^{B_0'} - 1] \tag{13.3}$$

2. The Birch equation-of-state (B-EoS, also called *second order Birch-Murnaghan EoS*):

$$P = \frac{3}{2}B_0[(\frac{V_0}{V})^{7/3} - (\frac{V_0}{V})^{5/3}] \tag{13.4}$$

3. The Birch-Murnaghan equation-of-state (BM-EoS, also called *third order Birch-Murnaghan EoS*):

$$P = \frac{3}{2}B_0[(\frac{V_0}{V})^{7/3} - (\frac{V_0}{V})^{5/3}][1 - \frac{3}{4}(4 - B_0')((\frac{V_0}{V})^{2/3} - 1)] \tag{13.5}$$

Note that the BM-EoS is identical to the B-EoS (13.4) for $B_0'=4$.

4. The Vinet ("universal") equation-of-state which incorporates explicitly also the temperature [118]:

$$P(V,T) = \frac{3B_0(T_R)}{X^2}(1 - X)exp[\eta_0(T_R)(1 - X)] + \alpha_0(T_R)B_0(T_R)(T - T_R) \tag{13.6}$$

where

$$X = [V/V_0(T_R)]^{1/3} \tag{13.7}$$

$$\eta_0(T_R) = 3/2[B_0'(T_R) - 1] \tag{13.8}$$

and T_R is a reference temperature, usually 300 K, and α_0 the volumic thermal expansion coefficient at zero pressure.

There are many other equations-of-state and considerable literature and discussion on which one is more appropriate. Some of them are semi-empirical, others can be derived from an interatomic pair potential, see Poirier [2] for an overview. The difference between them becomes however only appreciable at strong compression which are irrelevant for neutron scattering.

As for the expressions 13.3-13.6 it is important to realize that they are practically indistinguishable for small compressions, i.e. $\Delta V/V_0 < 30\%$. For example for $V/V_0-0.9$ and with $B'_0=4$, the pressure derived by the Murnaghan EoS and the Birch EoS differs by less than 1%. Since this is typically the compression in neutron scattering experiments, *the simple Murnaghan EoS (equation 13.3) is in most cases sufficient, even if the parameters are taken from one of the other equations-of-state.*

The M-EoS, B-EoS and BM-EoS do not explicitly contain a temperature dependence. It can be incorporated in first approximation by making V_0 temperature dependent $(V_0=V_0(T))$ through the known thermal expansion coefficient, and assuming B_0 and B'_0 temperature independent. This is in fact a reasonable assumption since it is known that B_0 and B'_0 vary only a few % in the temperature range between 0 and 300 K [2, 119]. In a further refinement of this model, also B_0 and B'_0 can be made temperature dependent. But this is often based on empirically rules since experimental data is rather scarce. Within the Vinet formalism the temperature dependence $B_0(T)$ and $B'_0(T)$ can be calculated from the ambient temperature values B_0, B'_0 and V_0:

$$B_0(T) = \frac{B_0}{X^2}[2 + (\eta - 1)X - \eta X^2]exp[\eta(1 - X)] \qquad (13.9)$$

$$B'_0(T) = \frac{4 + (3\eta - 1)X + \eta(\eta - 1)X^2 - \eta^2 X^3}{3[2 + (\eta - 1)X - \eta X^2]} \qquad (13.10)$$

where $\eta=3/2[B'_0-1]$. The temperature dependence enters hence via $X=[V(T)/V_0]^{1/3}$, i.e. the thermal expansion.

To derive pressures from the measured volume $V(P,T)$ and the parameters given below, the correct choice of $V_0(T)$ is critical. One should, if possible, not rely on literature values but measure it on the same instrument to avoid errors due to an inaccurate calibration.

13.2   Pressure markers

In this section we give a list of pressure markers useful for neutron diffraction when the sample itself cannot be used as pressure marker. At this point it should be mentioned that the use of pressure markers is not without pitfalls. The reason for this is the so-called Lamé effect [120]. A well known example is a mixture of finely dispersed gold powder in NaCl. When this mixture is compressed at ambient temperature, the pressure indicated by the lattice parameters of gold is up to 20% larger than that of NaCl, although the respective equations-of-state are accurately known! The difference disappears only when the mixture is heated to $\approx 300°$C, but reappears after further compression at 300 K. To understand this phenomenon consider a spherical inclusion in a spherical matrix which is subjected to a hydrostatic pressure at the outside. The bulk and shear moduli of the two materials are denoted by B_i and μ_i (inclusion) as well as B_m and μ_m (matrix). Then, the pressure in the inclusion is given by [120]:

$$P_i = P_o(1 + \frac{4\mu_m}{3B_m})/(1 + \frac{4\mu_m}{3B_i}) \tag{13.11}$$

The inclusion will hence feel a pressure difference which depends on the shear modulus of the matrix and the difference in bulk modulus between matrix and inclusion. If the matrix has a low shear modulus ($\mu \approx 0$), $P_i = P_o$, and the same is true when the bulk moduli are similar ($B_i \approx B_m$). If the opposite is the case, such as in a NaCl/Au mixture, a measurement of the EoS of the inclusion based on the pressure indicated by the matrix (or vice versa) will be erroneous. A safe way to remove this problem is to heat the sample sufficiently after each pressure increase - an option which is often not feasible.

13.2.1 NaCl

Rocksalt is since a long time the favorite pressure marker for x-ray diffraction experiments and has acquired the status of a pressure standard. It is also useful for neutron diffraction but has the disadvantage that chlorine has a quite large absorption cross section (33.5 barn, see Appendix G) and that it reacts with or is dissolved by many materials investigated by neutrons, such as aqueous solutions and hydrous solids.

For the equation-of-state of NaCl as a function of pressure and temperature one refers usually to tabulated values $V/V_0(P,T)$ obtained by Decker from a thermodynamic model using a classical interatomic potential which was fitted to the available experimental data [121]. This "Decker-EoS" spans the temperature range 0-800°C and was revised by Brown, using more recent experimental input data [122]. The deviation of Brown's EoS from Decker's EoS is at most 3% at 10 GPa (with the former giving higher pressure values), and is believed to be accurate to within the same level of error. For practical purposes an analytical EoS is often more useful. A recommendable form is a Vinet-EoS with parameters at 300 K:

- $V_0 = 179.43$ Å^3; $B_0 = 23.5$ GPa; $B_0' = 5.35$; $\alpha_0 = 1.2 \ 10^{-4}$ K^{-1}

Vinet showed that this EoS reproduces Decker's calculated values [118]. The Decker EoS can be extended to low temperatures, as demonstrated by Skelton et al. [123] who gives tabulated V/V_0 values down to 0 K.

13.2.2   Pb

Lead is in many aspects an ideal pressure marker for neutron scattering. It is a strong neutron scatterer but has a negligible incoherent and absorption cross section, i.e. it is basically transparent to neutrons. It has a low bulk modulus and is hence sensitive to small pressure changes, and is also an excellent solid pressure transmitting medium due to its low shear strength. The slight inconvenience is that it transforms at 13 GPa from the ambient pressure bcc to the high pressure hcp phase. But the volume change is small (1%) and the EoS of the hcp phase is reasonably well known [124, 125, 126]. Also, lead tends to form relatively large single crystals which leads often to very textured powder diffraction patterns. Since lead is more and more used as a pressure transmitting medium for low-temperature measurements, there is a need for knowing the EoS over a wide temperature range from 0 K to ambient temperatures.

The so far most reliable data on B_0 and B_0' at ambient and low temperatures are from ultrasonic measurements at 295 and 195 K [127], which were extrapolated to 0 K by Fortes et al. [128]. More recently, Strässle et al. (unpublished) measured the EoS of lead by neutron diffraction in the 80-300 K range, deriving the pressures from Brown's EoS of NaCl [122], which was extended to low temperatures. These measurements show very good agreement with the extrapolated ultrasonic data. Based on these diffraction results, a simplified EoS can be adopted in form of a Murnaghan EoS (13.3) with input parameters:

- 295 K: V_0=121.24 Å^3; B_0= 40.9 GPa; B_0'=5.5

- 180 K: V_0=120.08 Å^3; B_0= 43.3 GPa; B_0'=5.5

- 80 K: V_0=119.22 Å^3; B_0= 45.4 GPa; B_0'=5.5

- 0 K: V_0=118.66 Å^3; B_0= 46.5 GPa; B_0'=5.5

Values for intermediate temperatures can be obtained by simple interpolation. With these data for B_0 and B_0', the calculated pressures agree in the 0-300 K/0-10 GPa range to within better than 1% (1% in the 150-300 K range and better below) with the extrapolated ultrasonic data. In practice, the largest error in pressure determination is introduced by an imprecision of the measured V(T) and V_0(T). Appendix I gives a list with V_0(T) values for Pb (and NaCl) which might be useful for this purpose.

13.2.3   MgO

MgO (periclase) is the favorite pressure marker in geosciences given its relevance for the Earth's mantle. Its practical advantage is that it is inert and stable over a huge temperature and pressure range. For neutron scattering it is quite adequate since both Mg and O are good scatterers (Appendix G) with small incoherent cross sections. A particular feature of MgO is that the scattering lengths of Mg and O are almost identical, which means that the reflections with odd Miller indices are almost wiped out. As a consequence, the diffraction pattern will only be weakly contaminated by the pressure marker, in general by only 3 sharp reflections. Its inconvenience is its high bulk modulus and the fact that it cannot be used as a solid pressure transmitting medium, contrary to NaCl and Pb. The EoS of MgO is very well known and the scattering in the parameters is relatively small. We chose the values determined by ultrasonic measurements and a Vinet EoS:

- 300 K: V_0=74.68 Å^3; B_0= 160 GPa; B'_0=4.2

This set of parameters was shown to be consistent with high P/T diffraction data in the 0-10 GPa and at least to 1300 K [129, 130].

13.2.4 Ice VII/VIII

These two ice forms are stable between 2 and 60 GPa and their EoS have been carefully investigated by x-ray and neutron diffraction. Both are believed to have the same EoS since the volume change at the VII-VIII transition (which occurs around 0°C) is less than 10^{-3}. The favorite phase is ice VII since it is cubic (Pn3m), compared to its hydrogen-ordered partner ice VIII which has a small tetragonal distortion (space group I4$_1$/amd). Ice VII is a very strong scatterer and even small amounts of it can be detected. Its interest in neutron scattering is that it sometimes occurs as contaminant, as a result of de-hydration of the sample, for example. In this case the pressure can be easily determined, as long as it is above 2 GPa. In principle, also other ice phases can be used as pressure marker, but their EoS is much less known and their scattering intensity is much smaller due to their more complex structures.

Besson et al. suggested a temperature dependent EoS based on a Birch-Murnaghan equation with input parameters for various temperatures which were derived from published V(P,T) data fitted to a Vinet EOS. These BM-EoS parameters are [131]:

- 300 K: V_0=12.45 cm^3/mole; B_0= 20.4 GPa; B'_0=4.70

- 265 K: V_0=12.38 cm^3/mole; B_0= 21.0 GPa; B'_0=4.66

- 100 K: V_0=12.06 cm^3/mole; B_0= 23.6 GPa; B'_0=4.48

- 0 K: V_0=11.98 cm^3/mole; B_0= 24.1 GPa; B'_0=4.44

13.3 Useful empirical rules

In the absence of any knowledge of the equation-of-state of a sample which could be used to derive the pressure from the measured unit cell volume, one can make use of some "rules-of-thumb" which are rather well established and which work surprisingly well despite their apparent simplicity.

13.3.1 Bulk modulus (B_0)

It has been realized a long time ago [132, 133], that within a given class of solids the bulk modulus at ambient pressure B_0 often varies inversely with the ambient pressure molar volume V_0 [2]

$$B_0 V_0 = constant \qquad (13.12)$$

This holds for many oxides with spinel, rutile, corundum, wurtzite, and rocksalt structures, the constant in equation 13.12 being different for each class [2]. For example for Al_2O_3 and Cr_2O_3 the product B_0V_0 is identical within the precision of the measurements despite their vastly different molar weights. The same is true for MgO and SrO, and for GeO_2 and SiO_2. A strategy for finding the bulk modulus of a solid is hence to look for a system in the same class of solids (and the same structure) for which the bulk modulus B_0 is known and apply equation 13.12 with the measured V_0. It is almost certain that this strategy will work for solids which are related by doping, or even when one element is completely substituted by another. To give an example, let us consider the series of fcc monochalcogenides LaS, LaSe, and LaTe for which reasonably accurate values of B_0 are known. Starting with values for LaSe (B=74 GPa, a=6.067 Å) [134] equation 13.12 would predict for LaS (a=5.852 Å) B_0=82.5 GPa and for LaTe (a=6.545 Å) B_0=59 GPa which is in rather good agreement with the measured values of B_0=83-89 GPa (LaS) and B_0=55-61 GPa (LaTe) [134].

It is amazing how far the correlation 13.12 can be generalized, i.e. how the restriction to solids "with the same crystal structure" can be relaxed, and only the requirement of "similar type of binding" is retained. It appears therefore that volume is the predominant factor governing the bulk modulus, and structural details play in general only a secondary role. Glasser [135] generalized formula 13.12 to

$$B_0 V_0 \left(\frac{2}{n}\right) = constant \qquad (13.13)$$

where n is the number of atoms per formula unit. The factor 2/n converts V_0 to the molar volume of 2 atoms. With this definition, formula 13.13

---

[2]Note that the Murnaghan EoS (equation 13.3) would predict $B/B_0 = (V_0/V)^{B_0'}$ for pressure/volume changes *within the same phase*.

holds for certain classes of solids, irrespective if they have the same structure or not: the 1:1 monoxides, mono-chalcogenides, monopnictides, and chalgogenides have all the same constant in 13.13. Oxides with closely packed oxygen lattices (ex.: Al_2O_3), garnets (ex.: $Y_3Fe_5O_{12}$), and spinels (ex.: $MgAl_2O_4$) have a different common constant, as do the alkali halides.

To illustrate this concept we take the example of La_3S_4 [135]. Here n=7 and V_0=109 cm^3/mol, hence $2V_0/n = 31.14$ cm^3/mole. To estimate its bulk modulus we take the mono-chalcogenide reference system LaS which has similar binding, and for which B_0 is known: B_0=83 GPa [134]. With V_0=30.21 cm^3/mol = $2V_0/n$, the estimated bulk modulus of La_3S_4 is therefore (30.21/31.14)*83.3 GPa = 80.8 GPa. This deviates from the measured value B_0=74 GPa by 8% [135].

13.3.2 B-prime (B_0')

Most of the equations-of-state mentioned above require knowledge of B_0', the pressure derivative of B_0 at ambient pressure. This number is much more difficult to determine experimentally and the numbers quoted in the literature should be considered with much caution. Uncertainties are typically ±0.5, and the values derived from fits to compression curves are usually strongly correlated with the values of B_0. General observations are:

- For metals, B_0' spans a range of about 3 to 6.5 [136].

- Van der Waals solids have usually B_0'>7, some highly anisotropic systems like graphite can have B_0' as high as 12.

- Values B_0'<3 are rare and signal usually some electronic transition.

- The vast majority of covalently bonded solids have $4 \leq B_0' \leq 5$. Close-packed minerals fall into this category [2].

- Hydrogen-bonded solids have $4 \leq B_0' \leq 7$, with B_0' increasing with decreasing strength of the H-bond. Example: ice VII: B_0'=4-5 [131] whereas ammonia IV B_0'=6-7 [137].

These observations can be rationalized by the fact that B_0' is related to the form of the interatomic potential. For a Mie-potential

$$E(r) = \frac{-a}{r^m} + \frac{b}{r^n} \tag{13.14}$$

it can be shown that B_0' is given by [2]:

$$B_0' = \frac{1}{6}(n + m + 6) \tag{13.15}$$

For solids which can be described by a 6-12 Lennard-Jones potential (m=12 and n=6 in equation 13.14) one would therefore expect B_0'=8. This is indeed found in typical Van der Waals solids like neon [138] and argon [139].

To derive pressures from the measured change in unit cell volume one might therefore take an approximate EoS by "guessing" the value of B'$_o$ from the rules given below.

13.1 Pressure transmitting media (PTM)

The vast majority of high pressure experiments are aimed to be carried out under hydrostatic conditions. The reason is that hydrostatic pressure is a thermodynamic parameter and the results obtained under such conditions are intrinsic material properties which can be compared to theory. In practice this is realized by immersing the sample in a pressure transmitting medium (PTM), in the ideal case a fluid, but at sufficiently high pressures or low temperatures unavoidably a solid. Since the melting line of fluids eventually increases under pressure, solidification inevitably occurs at some pressure. Beyond this point, the pressure across the experimental volume is generally inhomogeneous and differential (mostly uni-axial) stress and shear stresses appear. Depending on the type of measurement, this leads to a more or less dramatic decrease of the quality and accuracy of the data and often to the appearance of "anomalies" which might be wrongly ascribed to new physical phenomena, see K. Takemura [140, 141] for a few illustrative examples.

The choice of the right PTM is hence critical since it limits the interpretation of the data. For this reason there is already quite a lot of literature on this issue [4, 5] and most of what has been said also applies to high pressure neutron scattering. Requirements on PTMs in this domain are specifically:

1. For liquids: high melting pressure, low melting temperature. For solids: low shear strength which is only weakly pressure dependent.

2. Small neutron cross section (absorption, incoherent and coherent), in particular no hydrogen.

3. Chemical inertness, if possible even to high temperatures, to avoid reaction with the sample.

4. No risk of neutron activation.

5. Low compressibility to avoid excessive volume changes.

6. For solids: simple or fully disordered structure to avoid Bragg reflections.

7. Cheap and readily available.

8. Easy to handle.

A particular difficulty is the requirement for media which contain no hydrogen which reduces considerably the number of substances which are at the same time cheap and readily available. For example the whole class of hydrocarbons which have been extensively used in the several-GPa range are excluded, the classical iso/n-pentane mixtures are almost impossible to obtain at reasonable cost.

Even if the pressure transmitting fluid does not contain hydrogen, its neutron scattering power is usually in the same order of magnitude as the sample. A diffraction pattern will therefore contain the signal of the sample superimposed over the broad oscillations produced by the fluid or glassy PTM. Figure 13.1 gives a typical example obtained on a sample of hematite (α-Fe$_2$O$_3$) in two different PTMs, the 4:1 ME and a 3:1 glycerine-water mixture. For well-crystallized samples the contribution of the PTM can be considered as background and poses no problem for the Rietveld refinements. But in high pressure studies of disordered materials the use of fluid or glassy PTMs is much more inconvenient. In this case, the best and possibly only choice is to use no PTM at all.

Another particularity is that most of the science encountered in neutron scattering involves low temperatures, often very low temperatures to below 4 K, which means any PTM will be solid at relatively moderate pressures. Whilst this is not avoidable, the effect of non-hydrostatic conditions can be limited by the appropriate choice of PTM.

What we give below is a selection of PTMs which are compatible with neutron scattering measurements and which have been well characterized. The ultimate choice depends on the P/T range of the experiment, the chemical nature of the sample, and the sensitivity of the physical quantity investigated to non-hydrostatic pressure conditions.

13.4.1   Fluorinert

Fluorinert is a trademark for a series of perfluoro-carbon liquids commercialized by 3M. Fluorinert liquids have been applied for a long time as PTM for neutron scattering [33] since they contain no hydrogen, the most popular types being FC75 and FC77 with generic chemical compositions $(C_8F_{18})_n(C_8F_{16}O)_m$. The hydrostatic pressure range is very limited, up to approximately 2 GPa, with a mixture of 1:1 FC84-FC87 being slightly superior [142, 143], i.e. a glass transition pressure of 2.3 GPa. Beyond, pressure gradients measured in a diamond anvil cell increase rapidly and reach 0.5 GPa at 10 GPa. The pressure-induced broadening of diffraction lines can be easily observed in medium-resolution ($\delta Q/Q \approx 1\%$) diffractometers such as PEARL. The conclusion is that Fluorinert is inadequate for pressures beyond 2 GPa, in particular for high resolution diffraction measurement or experiments on single crystals, both being highly sensitive to non-hydrostaticity.

13.4.2   4:1 Methanol-Ethanol (ME)

The 4:1 methanol-ethanol mixture is the most commonly used pressure transmitting medium and has been investigated by numerous groups which conclusively show that the glass transition at 295 K is at 10.5 GPa [4, 5]. Traditionally this mixture is used for DAC experiments, but since the advent of encapsulating gaskets (section 10.3) it is also routinely applied in high pressure neutron scattering to 10 GPa [92, 93]. Deuterated methanol and ethanol is easily available and the deuteration has a negligible effect on the solidification pressure [143]. The addition of a small amount of water does not increase the solidification pressure either, contrary to a widespread belief [143]. However, the hydrostatic pressure range can be considerably extended by only moderate heating. The gain is approximately 1 GPa per 11 K temperature increase. This is considerably larger than the slope of melting lines of pure molecular PTMs (example: hydrocarbons) which is typically 1 GPa per 70 K [5]. Heating to 100°C is possible with almost any pressure cell and will extend the hydrostatic pressure range up to 17 GPa. Even if the experiment is carried out at lower temperatures, this method allows to stay in the fluid phase at least during the compression phase where most of the non-hydrostatic stresses build up. At ambient pressure, ME freezes at 140 K from which the solidification ("melting") line can be obtained by interpolation, see Appendix E.

13.4.3   Other fluids

Osakabe and Kakurai investigated hydrogen-containing fluids for potential use in high pressure single crystal neutron scattering [144]. Since single crystal diffraction probes only a small solid angle around sharp Bragg reflections, the incoherent scattering from hydrogen can in many cases be tolerated. Using a triple axis spectrometer these authors investigated, among others, four hydrogen containing fluids: (A) Fomblin (a vacuum oil), (B) a silicon oil (KF96-50CS), (C) Daphne7373 (an olefin oligomer produced by Idemitsu Co., Ltd. Tokyo, Japan), and (D) glycerine ($C_3H_5(OH)_3$, also called glycerol). Daphne7373 is a well known PTM for medium pressure - low temperature research and solidifies at ambient temperature at 2.3 GPa [145, 146]. It is widely used for macroscopic low-temperature measurements using clamp devices where it is superior to other more common PTMs such as hydrocarbon fluids. Observation of the peak width Δ of a NaCl crystal compressed at 300 K by these media showed a rapid increase of Δ at 1 GPa or below in Fomblin and silicon oil, comparable to that in Fluorinert. A similar increase in width occurs in Daphne7373 only at 2 GPa, consistent with previous findings [146].

The case of glycerine is most interesting since it is reported to be quasi-hydrostatic up to at least 7 GPa with a measured peak broadening of only 50% over this pressure range [144]. Glycerine is a well-known glass-former which is used as a PTM since the early days of DAC research. Pure glycerine has a glass transition of 5.5 GPa at 300 K, as derived from ruby fluorescence and viscosity measurements [147]. The addition of water increases the hydrostatic pressure range [148], but this effect is rather modest: for a 3:2 (vol) glycerine-water mixture, which corresponds to the eutectic composition, the hydrostatic limit increases to 6.5 GPa [149]. Also, at this pressure most of the water crystallizes in the form of ice VII. This effect can be avoided by using a 3:1 (vol) mixture which does not produce any ice peaks. An important advantage of glycerine and glycerine-water mixtures compared to other PTMs is their low compressibility [150] (see Figure 13.2) and the smaller volume contraction upon cooling. These are favorable conditions for high pressure single-crystal studies where the pressure is changed in the fluid phase at high temperatures and the cell is cooled under approximately constant load through the glass transition. Glycerine is available in fully deuterated form at reasonable cost. Given all these advantages, it appears to be an adequate PTM for use in pressure devices in the 0-5 GPa range, in McWhan cells for example.

Figure 13.1 shows diffraction patterns of a medium-scattering sample, hematite (α-Fe$_2$O$_3$), collected in toroidal anvils at ambient pressure using the classical 4:1 ME mixture and a 3:1 glycerine-water (GW) mixture (all compounds fully deuterated). A pattern of the empty gasket in the same setup is given for comparison. The data show that in this scattering geometry more than half of the background is produced by the incoherent scattering of the TiZr gasket, the rest by the PTM. Both ME and 3:1 GW mixtures produce a broad feature between 20 and 30 degrees (Q=1.2-1.7 Å^{-1}) which is more pronounced in pure glycerine. Under pressure this feature moves rapidly to higher angles and flattens.

To resume, 3:1 glycerine-water and possibly Daphne appear to be favorable PTMs, especially for single crystal investigations. A new type of Daphne, Daphne7474, is available since recently which shows improved hydrostatic properties with a solidification pressure of 4 GPa at 300 K [151], and this fluid could have applications in high pressure neutron scattering. The potential advantage of these PTMs is that they are rather inert, rarely prone to dissolve samples, and that the size of the molecules is large enough to avoid penetration into the sample. Penetration of ME has been observed in open-framework structures like zeolites or nanoporous materials [152, 153], for example. Helium is known to penetrate into various phases of ice, as well as certain glasses [154].

13.4.4 Gases

Compressed gases, in particular noble gases, are unquestionably the best PTMs with He ranking top. There is considerable literature on this issue and the viscoelastic properties of He, Ar, Ne, and N$_2$ have been carefully characterized by diamond anvil techniques up to the 100 GPa range. To resume, in the 0-10 GPa range these gases have very low shear strengths, even in their solid phases and at low temperatures. Using ruby fluorescence techniques and compression at 300 K, the first detectable signs of non-hydrostaticity appear in He and Ne only at 15-20 GPa and in N$_2$ at 10-12 GPa, but in Ar as low as 2 GPa, i.e. immediately above the solidification at 1.4 GPa. For compression at low temperatures much less information is available, but it is clear that any other type of PTM, such as the 4:1 ME mixture, will produce stronger non-

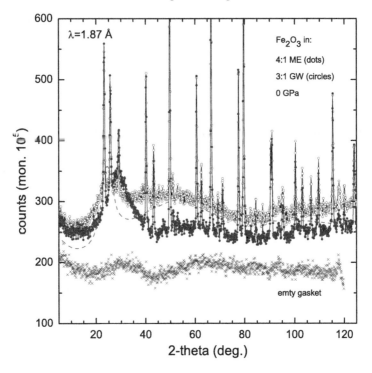

FIGURE 13.1

Background produced by two different PTMs, in comparison with the diffraction signal of a typical sample (hematite, α-Fe_2O_3) and of an empty TiZr gasket (crosses). Dots: hematite in a 4:1 ME mixture; circles: hematite in a 3:1 glycerine-water (GW) mixture. Data were collected at 0 GPa at the D20 diffractometer at ILL (Grenoble) in toroidal cBN anvils using TiZr gaskets in a 'longitudinal' scattering geometry as shown in Figure 10.5 (right). For comparison, the dashed line is the signal produced by pure glycerine.

hydrostatic stresses, even at low pressures. For low-temperature experiments compressed gases are clearly the PTM of choice.

The two obvious obstacles for their general use in neutron scattering are (a) the difficulty of loading at sufficiently high starting density, and (b) the strong volume change, even in the solid state. Figure 13.2 shows pressure induced volume changes $\Delta V/V$ for Ar, N_2, He and Ne and compares them to the 4:1 methanol-ethanol (ME) mixture [155] and glycerine. The starting densities were taken to be those of the fluids at ambient pressure. These densities are equivalent to the densities of the gases compressed at 300 K to 0.1-0.3 GPa, i.e. pressures within the reach of standard gas compressors. The plots in 13.2 are hence relevant for both cryogenic and high pressure loading.

Figure 13.2 illustrates the tremendous volume change in helium which

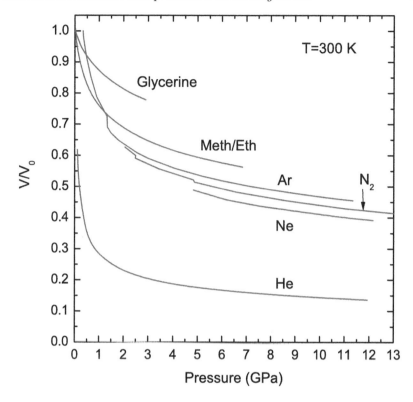

FIGURE 13.2
Variation of volume of various pressure transmitting media under pressure, at 300 K. For the gases V_0 corresponds to the specific volume of the liquid at ambient pressure, i.e. 34.63 cm^3/mol (N$_2$), 32.03 cm^3/mol (He), 28.68 cm^3/mol (Ar), and 16.72 cm^3/mol (Ne). Data taken from Refs. [147] (glycerine), [155] (ME), [156] (N$_2$), [157] (He), and [158] (Ar).

solidifies at 300 K only at 10 GPa [159]. The volume change between the density of the liquid at 1 bar and 10 GPa is a factor 7. Even if the loading can be achieved, the gasket and anvil-assembly has to be able to support such a volume reduction. For example the single toroidal anvils with the standard cup depth can only produce a volume reduction by a factor 3 at most, i.e. even with a large initial gasket thickness.

The situation is better with Ar which is easier to load (it can be done cryogenically) and where the volume reduction is only a factor 3-4 to 10 GPa. Argon has also the advantage that the natural isotopic composition is a weak neutron scatterer. Since its equation of state is well known, even at low temperature, it can be used as a pressure marker. To give an idea relevant for neutron scattering, Argon was used to compress mm-size single crystals of

NaCl and squaric acid ($H_2C_2O_4$) to 15 GPa without any detectable broadening of the Bragg reflections [114].

Nitrogen is potentially the most interesting PTM for neutron scattering. It is easy to load, inexpensive, and its volume change between the liquid and 10 GPa is only a factor 2, not very different from the ME mixture (Figure 13.2). Ruby fluorescence measurements in a DAC indicate no observable pressure gradients up to 12 GPa, i.e. visco-elastic properties much better than argon [143]. The reason for this is probably the presence of a weakly first-order phase transition from the hexagonal β to the cubic δ phase at 4.5 GPa (Figure 13.2) and the fact that both are plastic phases with the N_2 molecules dynamically disordered. Since 10 GPa is typically the pressure range exploited by neutron scattering, nitrogen will gain importance as a PTM for low-temperature work in this domain.

13.4.5   Solid PTMs

The central quantity to evaluate the performance of a solid PTM is its *shear strength* τ. Similar to yield strength measurements, shear strengths are determined by measuring the yield of a test rod under torsion, i.e. applying a torque and determining the point where yield occurs. Measurements of the shear strength of a material under pressure can be done by subjecting a disc-shaped sample under pressure using Bridgman anvils and measuring the required torque when one of the anvils is rotated. The important message is that this quantity is a measure of the maximum differential stress a solid, in particular a PTM, can withstand. Given a shear strength τ, the maximum stress that can occur at a given point of a material is given by Tresca's plasticity criterion:

$$|\sigma_i - \sigma_j| = 2\tau \qquad (13.16)$$

where σ_i and σ_j are any combination of the three principal stresses of the diagonalized (3x3) stress tensor σ. Recall that for hydrostatic pressure P all principal stresses are equal with $\sigma_1=\sigma_2=\sigma_3=P/3$. Figure 13.3 gives shear strengths of a series of solid PTMs at 300 K as quoted by Sherman and Stadtmuller [5] and which are a priori compatible with neutron applications. Solids which activate under neutron radiation (Cs, Rb, I) or which are strong neutron absorbers (BN, In) were not included. Note the very small shear strength of lead and its weak pressure dependence compared to NaCl.

The use of solid PTMs has advantages for low temperature high-P measurements since their shear strengths, though elevated compared to a fluid, remains smaller than many organic fluids once they are frozen. The archetypal ME mixture for example is likely to have very large shear strengths below its freezing point at 130 K (at ambient pressure).

The temperature dependence of shear strengths $\tau(T)$ has been shown to follow closely an exponential relation [5]:

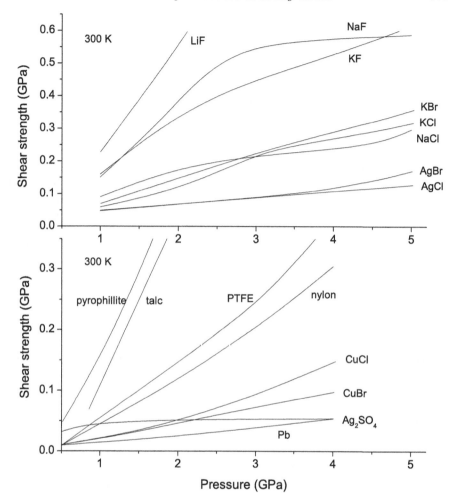

FIGURE 13.3
Pressure dependence of shear strengths of various solid PTMs, selected from
Ref. [5]. Related compounds containing Cs, Rb, I and In were not included
due to their risk of neutron activation or strong neutron absorption.

$$\tau(T) = \tau_0 exp(-\alpha T/T_m(P,T) \tag{13.17}$$

with some adjustable parameters τ_0 and α, and $T_m(P,T)$ the melting temperature at pressure P. It implies hence that pressure gradients can be removed
by heating close to the melting temperature, but also that shear strengths
increase considerably at low temperatures. The data on lead suggest for example that its shear strength at ambient pressure increases by two orders of

magnitude at 100 K [5]. From equation 13.17 it is clear that any change of pressure at low temperatures is prone to producing differential stress, and a recommended procedure is to change pressures at temperatures as high as possible and cool under more or less isochoric conditions.

Among the solids mentioned in Figure 13.3 the case of AgCl should be mentioned. Its shear strength follow relation 13.17 but with a weak temperature dependence, and it appears to be the best non-metallic solid PTM. Unfortunately both Ag and Cl are neutron absorbers, but in certain cases this inconvenience might be tolerable.

The pressure distribution within toroidal anvils has been measured for the case of NaCl as PTM [160], see Figure 13.4. With the center of the pellet at 6.0 GPa, the maximum pressure gradients occur in the direction of compression with values of typically 0.1 GPa/mm, and the maximal difference within the pellet is approximately 0.4 GPa, hence ±3-4%. It can be shown that in flat Bridgman-type anvils (and DACs), the maximum pressure gradient in radial direction is related to the shear strength by:

$$\tau = \frac{h}{2}\frac{\partial p}{\partial r} \tag{13.18}$$

where h is the thickness of the sample. Applying equation 13.18, Maede and Jeanloz find τ=0.22 GPa at 6 GPa [161], a value in reasonable agreement with the value indicated in Fig. 13.3. Although this relation cannot be strictly applied to Bridgman anvils with cups, taking an average sample thickness of h=4 mm and a measured pressure gradient of 0.1 GPa/mm, equation 13.18 would predict a shear strength of 0.2 GPa, which is consistent with previously cited values found in the literature. With this in mind and comparing shear strengths as given in Figure 13.3 one can safely predict that at 6 GPa the pressure variation for Pb and Ag_2SO_4 would be less than 0.2 GPa across the entire pressure chamber, for the same geometry. In reality the visible sample is located close to the equatorial plane where the pressure gradients are much smaller.

There exists indeed considerable experience in the use of lead as PTM for high pressure single crystal studies in inelastic neutron scattering, including at low temperatures [162, 163]. The procedure to embed a 20 mm^3 (diam. 4 mm, height: 1.5 mm) oriented single crystal in a lead matrix surrounded by the gasket is shown in Figure 13.5. Such experiments use toroidal anvils with a conoidal cup, a geometry which favors maintaining the orientation of the single crystal during the preparation of the pellet and during the compression in the experiment. The dimensions of the corresponding anvils are given in Appendix D. A similar procedure might be applied to powder samples as well.

With this method single crystals of Ge, GaSb, Fe, and various oxides could be compressed up to 9 GPa [162]. The rocking-scans showed no pressure-induced increase of the mosaicity and these crystals could be recovered without damage after the measurement. In more fragile crystals like PbTe and Zn an increase of the mosaicity to approximately 1° could be observed. The

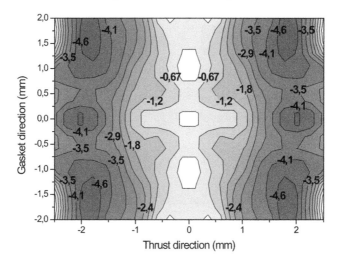

FIGURE 13.4
Pressure distribution in a NaCl pellet compressed by toroidal anvils similar to those shown in Figure 10.3. Numbers are deviations of the pressure in kbar from the center of the pellet which is at 6.0 GPa. Data were obtained using sintered diamond anvils and energy dispersive x-ray diffraction with a spatial resolution of 0.5 mm. Reprinted from Ref. [160].

relevant parameter for evaluating if this method risks to damage samples is a comparison of the shear strength of the sample with that of lead at a given pressure.

13.5 General comments and recommendations

Bridgman stated that "...the most important immediate problem of technique in this field is to find methods of producing stress systems which are truly hydrostatic" [164]. Half a century later this is still true, in particular for neutron scattering which deals largely with low-temperature phenomena (magnetism, quantum critical phenomena, hydrogen ordering) where true hydrostatic pressures are virtually impossible or difficult to achieve. This technical problem is amplified by the fact that magnetic ordering is often sensitive to non-hydrostatic stresses, a well documented case is the Morin transition in hematite Fe_2O_3 studied by high pressure neutron scattering using different PTMs [165]. The transition-metal monoxides such as NiO and FeO are other examples, where both the rhombohedral splitting and the intensities of the

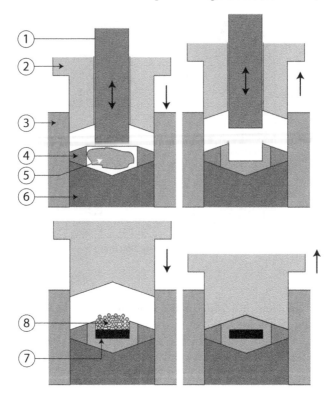

FIGURE 13.5
Procedure to prepare a gasket-sample pellet with a single crystal embedded in a Pb-matrix. (2)&(6) pistons; (1) adjustable shaft; (3) cylinder; (4) metallic gasket (diam. ext.: 10 mm, diam. int.: 6 mm); (5) lead pellet; (7) single crystal; (8) lead powder.

magnetic Bragg reflections depend highly on the stress state of the sample. It is therefore clear that high pressure measurements at low temperatures need particular care in both their preparation and the interpretation of the data, and in the long term it seems unavoidable to use compressed gases as PTM. The technical solutions for gas loading are discussed in chapter 12.

14

Applications

CONTENTS

14.1 General

This chapter presents a number of science examples obtained with equipment discussed previously. They are grouped in classes which identify issues and techniques which are typical - and maybe unique - for neutron scattering and which cover a large part of the community working in this area: Structure of magnetic and low-Z materials, lattice and spin dynamics (phonons, magnons), crystal field excitations, small-angle neutron scattering (SANS), and quasi-elastic neutron scattering (QENS). The aim of this chapter is to illustrate to the reader what kind of data can be obtained or expected under certain P/T conditions using certain high pressure devices, and a certain instrument. The emphasis in the choice of the examples was put on more recent high pressure methods which have emerged since the early 1990s, in particular sapphire anvil cells and Paris-Edinburgh type cells. The examples will hopefully demonstrate the technical issues related to a given high pressure device, its constrains in terms of background, accessible Q range, control of pressure and temperature, compatibility with high magnetic fields, and other experimental parameters. If adequate, examples of both powder and single crystal applications were chosen to demonstrate their often quite different requirements. There are certainly many other good examples in the literature which would merit to be cited in this chapter but which could not be included for reasons of space.

14.2   Diffraction

14.2.1   Magnetic structures

Magnetic structure determination is a core issue in neutron scattering and the application of high pressure is particularly interesting. The problem is that it involves usually other parameters such as rather low temperatures (at least liquid He temperatures) and in the ideal case also high magnetic fields. These additional constrains tend to deter the magnetism community from high pressure studies. However, the advantage is that magnetic phenomena are usually very sensitive to high pressure (volume changes) and that a few GPa are usually enough to observe massive changes in magnetism.

Figure 14.1 (left row) gives an example of powder diffraction patterns recorded in a sapphire anvil cell at the D-12 diffractometer (Dubna) [46] on a $YMnO_3$ manganite sample of 2.5 mm^3 volume. These data were obtained by time-of-flight measurements where the incident beam is along the thrust axis and the diffracted neutrons are detected at a fixed scattering angle of either 90° or 45°. No pressure transmitting medium was used in these measurements. The lowest temperatures which can be achieved with the associated cryogenic equipment (a closed cycle refrigerator) is 10 K. Note the low level of contamination from the pressure cell. Typical accumulation times per pressure point were 20 hours. $YMnO_3$, which crystallizes in a hexagonal structure of space group P6$_3$mc, is antiferromagnetic below 70 K (at 0 GPa), as seen on the occurrence of the (100) and (101) reflection. Below T_N the Mn^{3+} spins order in the hexagonal plane. Rietveld refinements of the spectra show that under pressure to 5 GPa the direction of the spins turns by almost 50°, while their magnitude decreases from 3.3 μ_B to 1.5 μ_B.

The same sample was also measured at the G6.1 diffractometer [167] of the LLB at the French ORPHEE reactor, this time in an angular dispersive mode with the incoming and diffracted beams in the equatorial plane of the sapphire anvils, see Figure 14.1 (right row). This mode of operation allows access to low-Q (large d) reflections which reveal the existence of magnetic diffuse scattering (inset Figure 14.1) over a wide temperature range. The vertical orientation of the cell allows the use of standard ("orange") He-cryostats to reach temperatures as low as 1.5 K. NaCl was used as pressure transmitting medium.

The majority of current applications of sapphire cells of the Kurchatov-type design concern powder diffraction measurements carried out without or with solid pressure transmitting media like NaCl. Single-crystal diffraction measurements under hydrostatic conditions are much less frequent, Figure 14.2 gives an example by Link et al. [168] on a single-crystalline sample of TmTe with a volume of 0.003 mm^3, compressed in a 4:1 ME fluid. TmTe is a divalent magnetic semiconductor with a gap of 0.35 eV which crystallizes in the NaCl structure and orders at ambient pressure antiferromagnetically

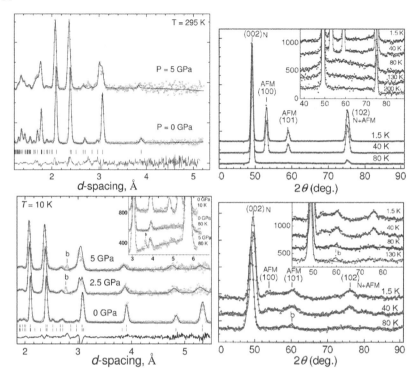

FIGURE 14.1

Powder diffraction patterns of YMnO$_3$ obtained with Kurchatov-Dubna sapphire cells. Left row: by time-of-flight measurements at P=0, 2.5 and 5 GPa and scattering angles 2θ=90 deg (main figures) and 45 deg (inset); T=290 K (upper) and 10 K (lower). At ambient pressure T_N=80 K and the AFM (100) and (101) reflections are located at 5.31 and 4.82 Å, respectively. Right row: by angle-dispersive diffraction with λ=4.74 Å, at 0 GPa (top) and 6.7 GPa (bottom). Note the diffuse scattering in the inset. "b" indicates the only visible signal from the pressure cell. Reprinted with permission from Refs. [53, 166].

(AFII-type) below 0.5 K. Under pressure the gap closes at 2 GPa and a mixed-valent state is observed beyond, without any structural transition. Figure 14.2 shows that this electronic transition entails ferromagnetic ordering as high as 14 K. Note that the complete solution of the magnetic structure requires also the magnitude of the order parameter (the magnetic moment) and the population of the different magnetic domains. The experiments show that the Curie temperature decreases under pressure and vanishes at approximately 7 GPa. The magnetic moment shows a similar trend and decreases from 2.1 μ_B at 2 GPa to below 1 μ_B at 5.5 GPa, which is considered the detection limit.

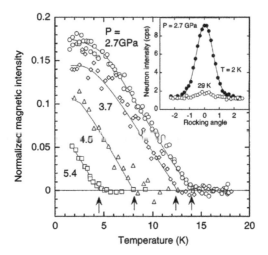

FIGURE 14.2

Single-crystal diffraction in a Kurchatov-LLB sapphire cell: Temperature dependence of of the magnetic contribution to the (111) peak intensity of TmTe, normalized to the (002) nuclear intensity, at different pressures. Arrows mark the Curie temperatures. Reprinted with permission from Ref. [168].

Certain investigations of magnetic phase diagrams are examples of neutron scattering at the highest pressures achieved so far, see Figure 14.3. These were carried out in a diamond anvil cell and sample volumes as small as 0.001 mm^3, at the G6.1 diffractometer of the LLB using wavelengths between 4 and 4.8 Å. This instrument is equipped with a two-stage NiTi supermirror "trumpet" which allows focalisation of neutrons which leads to a gain of intensity by a factor up to seven. Typical accumulation times were 12 h. The materials under investigation were the europium chalcogenides EuX (X=O,S,Se) and the gadolinium monopnictides GdX (X=As,Sb,Bi). These crystallize again in the NaCl structure and are considered model magnetic systems where the magnetic interactions can be tuned by high pressure. Both Eu^{2+} and Gd^{3+} have an immense magnetic moment of $7\mu_B$ which enable experiments with extremely small samples (note that the magnetic intensity varies with the square of the magnetic moment). Also, the strong absorption of natural Eu and Gd was reduced by the use of isotopically enriched ^{153}Eu and ^{160}Gd. At ambient pressure, the heavier compounds such as EuTe and GdAs of each family order antiferromagnetically. Figure 14.3 shows diffraction patterns for both compound up to 12 and 43 GPa. EuTe is found to transform at 9 GPa from an AF order with propagation vector k=(1/2,1/2,1/2) to one with k=(1/4,1/4,1/4), and to a ferromagnetic long-range order at 12 GPa. This remains up to 17 GPa where a structural transition to the CsCl (B1) structure occurs which wipes out magnetic order. In contrast, GdAs behaves different, the ferromagnetic structure is stable up to highest pressures which could be generated, i.e. 50 GPa.

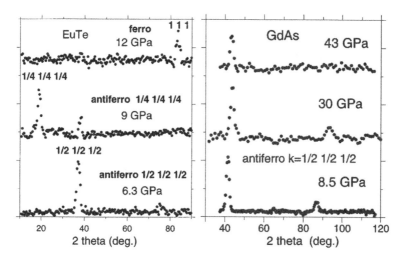

FIGURE 14.3

Powder diffraction patterns obtained in diamond anvil cells. Left: EuTe which transforms from an ambient pressure AF structure to ferromagnetic ordering at 12 GPa via an intermediate AF structure. Right: GdAs which remains ferromagnetically ordered up to 50 GPa. Reprinted with permission from Ref. [169].

It is clear that the use of diamond anvil cells to study magnetic structures by powder neutron scattering will remain restricted to few systems with exceptionally large magnetic moments. For general samples order of magnitude more sample volumes are required. For this purpose McWhan-type cells are still very useful and competitive since they offer cm^3 size sample chambers, relatively high pressures of up to 3 GPa and large neutron transparency. The contamination from the Al_2O_3 cylinder is usually no problem since the first Bragg peaks occur at much higher Q (smaller d-spacings) than the magnetic reflections.

Figure 14.4 gives an example of single-crystal work on the heavy-fermion compound URu_2Si_2 by Amitsuka et al. [170, 171] using a McWhan cell to 2.8 GPa, and a mixture of FC70 and FC77 Fluorinert as pressure transmitting fluid. In this system superconductivity at $T_c=1.2$ K was suggested to coexist with a weak antiferromagnetic ordering. The crystal structure is tetragonal, space group I4/mmm. Below 17.5 K the 5f moments of U order with a tiny saturation moment of 0.03 μ_B/U. Under pressure the intensity of the (100) peak increases dramatically which suggests an increase of the magnetic moment to 0.4 μ_B/U at 1.5 GPa. The same increase was later also observed in a CuBe clamp cell to 1.2 GPa [172] which allowed to measure ac-susceptibility and hence to detect superconductivity using the same setup. From these combined magnetic measurements it was concluded that the increase in magnetic moment is in fact correlated with a decrease in the superconducting volume fraction which casts strong doubt on the conjecture of coexistence of AF order with superconductivity.

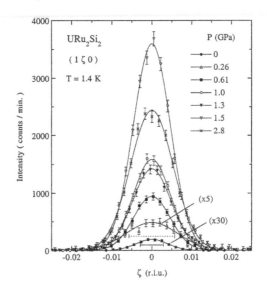

FIGURE 14.4
Single-crystal diffraction with a McWhan-type cell: Transverse scan across the (100) magnetic reflection of URu_2Si_2, collected at 1.4 K at different pressures. Reprinted with permission from Ref. [171].

Another interesting example for the use of CuBe clamp cells is given in Figure 14.5. In this case the pressure cell was placed into a high magnetic field of up to 6.5 T during the experiments. Diffraction measurements were carried out on the E4 diffractometer of the Hahn-Meitner-Institute (HMI, Berlin) with a wavelength λ=2.44 Å and a single crystal of Y_2Fe_{17} embedded in mineral oil with its a-axis and the rotation axis of the spectrometer parallel to the magnetic field. Y_2Fe_{17} crystallizes in a hexagonal structure and shows an in-plane collinear ferromagnetic order of the Fe-moments below its Curie temperature at 310 K. Under pressure of 8 kbar, superstructure reflections appear near the (002) (Figure 14.5), (004), and (300) nuclear peaks which show a strongly temperature dependent intensity. These observations are evidence for an incommensurable helimagnetic structure, reminiscent to what is found in Ce_2Fe_{17}, Tm_2Fe_{17}, and Lu_2Fe_{17}. The structure can be described as ferromagnetic planes rotating by an angle α along the c-axis. At 5 K $\alpha \approx 19°$ and the magnetic periodicity is approximately 4.65c.

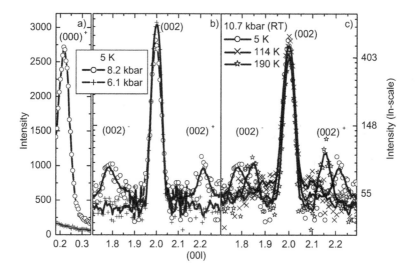

FIGURE 14.5

Single-crystal diffraction patterns collected with a CuBe clamp cell: Pressure-induced helimagnetism in Y_2Fe_{17}. Panels (a) and (b): (00l) scans at 5 K and 6.1 and 8.2 kbar. (c): temperature dependence of the pressure-induced $(002)^\pm$ satellites. Reprinted with permission from Ref. [173].

The observation of weak superstructure reflections is not restricted to single-crystal diffraction. Figure 14.6 gives an example of neutron powder diffraction patterns collected in a TiZr pressure cell using He as pressure medium [174]. The compounds of interest are Ce_2Fe_{17} and $Ce_2Fe_{16}Mn$, materials related to what is shown in the single-crystal study of Figure 14.5. Ce_2Fe_{17} is a mixed-valent compound which orders ferromagnetically below $T_C=94$ K and antiferromagnetically between $T_N=206$ K and 94 K. $Ce_2Fe_{16}Mn$ is AF-ordered below $T_N=200$ K with an incommensurate helical magnetic structure where the moment direction turns by 136 deg in going through one lattice (c) parameter. Under pressure, ferromagnetic order disappears in Ce_2Fe_{17} for P > 2.5 kbar and, on the contrary, is restored in $Ce_2Fe_{16}Mn$ at P > 1.5 kbar. Figure 14.6 shows diffraction patterns of both compounds obtained at the two-axis diffractometer D2B (ILL, Grenoble) under pressure, at a wavelength of $\lambda=2.52$ Å. Note the relatively high background due to the TiZr pressure cell (not the He pressure transmitting medium which is frozen at 3 K and 4 kbar).

If relatively large sample volumes (> 10 mm^3) and simultaneously high pressure (> 3 GPa) is required then a possible - maybe the only - option is the

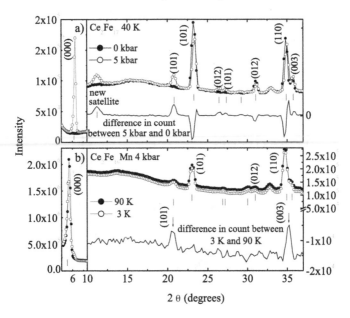

FIGURE 14.6
Powder neutron diffraction patterns obtained with a TiZr gas cell: (a): Ce_2Fe_{17}; (b): $Ce_2Fe_{16}Mn$. Reprinted with permission from Ref. [174].

PE press. With the cryogenic equipment discussed in chapter 11 temperatures down to 3.6 K can be routinely achieved[1] which enables magnetic structure determinations in the multi-GPa range. A clear disadvantage remains the size of the cell which excludes the application of high magnetic fields.

Figure 14.7 shows diffraction patterns obtained in a PE press at 7 GPa, on solid oxygen phase δ, stable between 6 and 8 GPa, and 0-300 K (approximately). The O_2 molecule has a magnetic moment of 1.6 μ_B which is able to produce long-range order in solid oxygen. This occurs at low temperatures (< 24 K at ambient pressure), but also in the high pressure δ-phase. δ-O_2 is tetragonal (Fmmm, Z=4) with the O_2 arranged in layers where the nearest in-plane O_2-O_2 distance is typically 2.6 Å and the nearest inter-layer distance is 3.7 Å. Magnetism in δ-O_2 (as well as the ambient pressure α-phase) is hence essentially two-dimensional. Figure 14.7 shows that there are in fact three different magnetic structures in the same nuclear structure, which differ in the sequence of antiferromagnetic order in the O_2-planes along the c-axis. Given a sample volume of 70 mm^3 and the high flux of D20, the accumulation times per refinable spectrum for such samples were as short as 20 minutes which allowed to map the entire magnetic phase diagram of δ-O_2 in half a day. A

---

[1] Temperatures of 2.1 K were reached by filling the inner chamber with liquid He and pumping on it (Th. Strässle, priv. comm.).

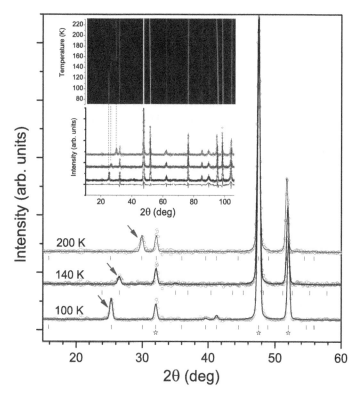

FIGURE 14.7
Powder diffraction patterns obtained with a PE-cell: Solid oxygen phase δ at
7 GPa and various temperatures. The occurrence of three different magnetic
structures within the same nuclear structure is visible by the magnetic reflections (arrows). Inset: Full spectrum with Rietveld fit (lower) and map of 32
patterns over a T-range of 160 K. Adapted from Ref. [112].

key feature of this method is to have sufficient intensity to work in a high-
resolution mode which allows to determine both the nuclear and the magnetic
structure simultaneously. This is crucial in the case of oxygen where there is a
strong interplay between nuclear structure and magnetic degrees of freedom.

Figure 14.8 illustrates an example of uni-axial pressure measurements for
the study of magnetic ordering in CeSb [175]. This compound crystallizes at
ambient conditions in the NaCl structure and exhibits at low temperatures
antiferromagnetic ordering of the Ce-moments in various phases. Each of them
can be characterized by a periodic stacking of ferromagnetic (001) layers with
'+' and '-' magnetization and non-magnetic '0' layers giving rise to an ordering
wave vector $q_0 = 2\pi/a(0,0,q_0)$, see the phase diagram in Figure 14.8. The high
temperature phase (just below $T_N = 14.5$ K) has a stacking sequence "+ - 0",

hence $q_0=2/3$. The interesting phenomenon in this system is the existence of a critical end point at p=2.0 kbar and T=18 K, where the paramagnetic and two antiferromagnetic phases with $q_0=1$ (AF-I) and $q_0=2/3$ (AF-IA) meet, see the phase diagram in Figure 14.8. The experiment was carried out with a single crystal of 3x3x3 mm dimensions and the device shown in Figure 9.3 which allowed p-T scans 'A' and 'B' close to the critical end point, as indicated in Figure 14.8 (lower right panel). Data were collected on a double-axis diffractometer of the former SAPHIR reactor at Würenlingen (Switzerland). These measurements detected Bragg reflections associated with the two magnetic structures (upper and lower left panels) as well as magnetic fluctuations (upper right panel) associated with the critical en point.

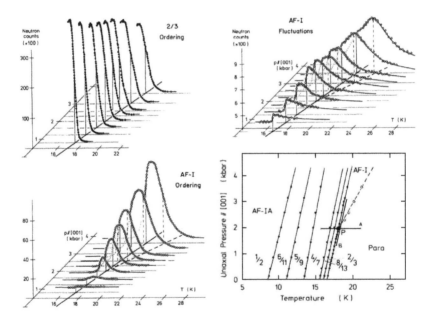

FIGURE 14.8

Single-crystal diffraction under uni-axial pressure: Magnetism in CeSb. Panel lower-right: Phase diagram as a function of temperature and stress along [001], with the critical endpoint P at 18 K and 2 kbar. 'A' and 'B' indicate scans shown in the other panels. See text for details. Reproduced with permission from Ref. [175].

14.2.2   Nuclear structures

This concerns mainly structural studies on low-Z material, oxides, and hydrogen-rich materials. There are also some issues involving heavier atoms which cannot easily be distinguished by x-rays, such as cation ordering phenomena in minerals, where neutrons have clear advantages. Most of the studies concern powder samples where pressure-induced phase transitions are involved which a single crystal would in most cases not survive. As a result of this, powder diffraction is in most cases the only option for high pressure structural studies.

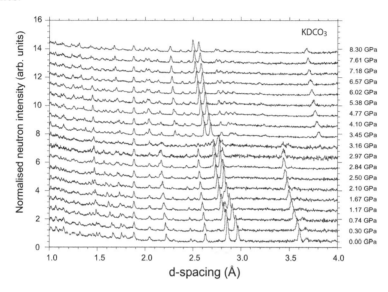

FIGURE 14.9
Powder diffraction using a PE-cell: Sequence of neutron diffraction patterns of $KDCO_3$ as a function of pressure through a phase transition at 3 GPa. Reprinted with permission from Ref. [176].

In Figures 14.9 and 14.10 an example of a high pressure study on kalicinite (potassium bicarbonate, $KHCO_3$) is given, carried out on powder samples in a PE press at the PEARL/HiPr beamline of the ISIS facility [176]. Under ambient pressure $KHCO_3$ crystallized in the monoclinic $P2_1/a$ space group. Its structure is characterized by the presence of strongly bonded $(HCO_3^-)_2$ dimers, short $O\cdots O$ and $O\cdots H$ distances of of 2.656 Å and 1.64 Å, respectively, and a slightly bent $O\cdots H$-O arrangement with an angle of 176°. The hydrogen-bond can therefore be classified as unusually strong. Under pressure a first-order phase transition occurs at 3 GPa which is easily visible in the raw data shown in Figure 14.9. Figure 14.10 shows diffraction patterns of the two phases in more detail, together with a Rietveld fit. Note that these include reflections from the TC anvils (WC and Ni) as well as the lead pressure marker. Using these patterns combined with high pressure x-ray data the structure of the high-pressure phase could be solved in space group $P\bar{1}$ and full structural data obtained up to 8 GPa. These indicate that the structural motives remain largely preserved and that the hydrogen bond-angle O··· H-O bond-angle decreases to 165°, which can be interpreted as a weakening of the hydrogen-bond.

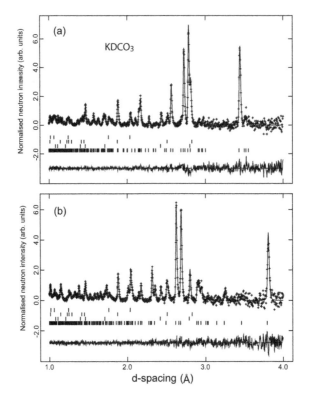

FIGURE 14.10
Diffraction pattern of $KDCO_3$ at 2.8 GPa (a) and 3.4 GPa (b) in the $P2_1/a$ and the $P\bar{1}$ structures, respectively. The line is a Rietveld fit to the data, tickmarks in the bottom line are those of the sample, others are from the tungsten carbide anvils (WC and Ni) and the Pb pressure marker. Reprinted with permission from Ref. [176].

The quality of powder pattern collected at PEARL/HiPr in TC anvils in the 0-10 GPa range as illustrated in Figure 14.10 may be compared with the one obtained in sintered diamond (Co-SD) anvils under pressure above 20 GPa, which allow only sample volumes of $\approx$ 30 mm^3. Figure 14.11 shows diffraction patterns of ammonia phase IV at 6.5 GPa and 23 GPa. The structure of this phase in the 4-9 GPa range (space group $P2_12_12_1$, Z=4) has been solved previously by Loveday et al. [177] using data obtained at the former POLARIS station, also at ISIS, in TC anvils. In both cases the sample was loaded by condensing the gas at $\approx$ -5°C into the anvils using a procedure which is explained in Figure 12.2. Although there is no evident phase transition visible in the patterns of Figure 14.11 the Rietveld refinements reveal a subtle distortions of the ND_3 molecule occurring above 20 GPa leading to an almost straight N$\cdots$H-N hydrogen bond along the z-axis. Evidence for such pressure-induced distortions is also visible in the lattice compressibility obtained by x-ray diffraction, and in the change in Raman spectra beyond 20 GPa.

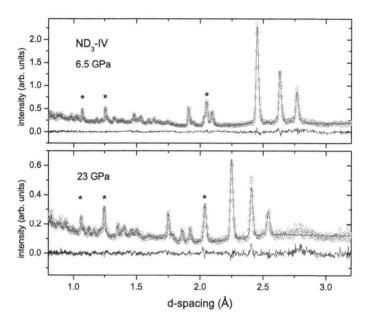

FIGURE 14.11
Powder diffraction with sintered diamond anvils: ammonia ND_3 phase IV, collected at the PEARL/HiPr station, ISIS. The lines through the data (circles) are Rietveld fits. Asterisks (*) indicate reflections from diamond. Adapted from Ref. [110].

Obtaining the most detailed high pressure structural information requires diffraction on single-crystals. Typical issues are multisite disorder of hydrogen or diffuse scattering. Such experiments need highly hydrostatic pressure conditions and panoramic pressure cells to access a maximum amount of reciprocal space, as well as accurate absorption corrections.

There have been various successful experiments using McWhan cells as well as sapphire anvil cells for single-crystal studies at pressures below 3 GPa. An example of one of the most detailed single-crystal diffraction study carried out in a McWhan cell (the Bloch-type cell described in Ref. [32]) is the work by Tibballs et al. [42] on KH_2PO_4 (KDP) and KD_2PO_4 (DKDP) as a function of pressure and temperature. Cylindrical crystals of 3 mm diameter and 3 mm height were compressed at room temperature to 1.7 GPa using Fluorinert and also cooled to 80 K. The data quality was good, with no significant background problems.

A similar scientific problem was later studied on $PbDPO_4$ in a sapphire cell (of the Ahsbahs-type) to 1.4 GPa [178] compressed in Fluorinert PTM, in the temperature range T=230-330 K, also carried out at ILL. The advantage of this technique is the large panoramic view, the obvious disadvantage the smaller sample volume ($\approx$ 0.4 mm^3). A version with sapphire sphere anvils was later used to study again KH_2PO_4 to 2.9 GPa [51], on a smaller sample (0.8 mm^3) and using a 4:1 ME mixture as PTM. Another example for single-crystal diffraction using this type of cell is the study on ice VI to 0.9 GPa [52].

We will discuss in more detail a few examples of single-crystal diffraction in the 10 GPa range and beyond which illustrate the potential but also the technical challenges related to it.

The first example is a single-crystal diffraction study to 10 GPa on squaric acid ($H_2C_4O_4$) using a PE-press and the D9 diffractometer of the ILL [179], see Figure 14.12. $H_2C_4O_4$ is tetragonal with C_4O_4 units in the a-b plane, linked by hydrogen bonds. It has been suggested that under high pressure the hydrogen bonds center. The experiments were carried out using a two-column VX-type press, toroidal SD anvils, single crystals of 3-4 mm^3 volume in a TiZr gasket and a 4:1 ME pressure transmitting fluid. The measurement requires an accurate and stable orientation of the sample. A considerable drawback is the limited azimuthal aperture of approximately $\pm$ 5° imposed by the anvils, and to a minor extend the shadowing of the two pillars. Access to a maximum of reciprocal space needs short wavelengths, hence a hot source, in this case wavelength between 0.33 and 0.71 Å. With this setup more than hundred reflections could be measured at 10 GPa, 57 of them independent. Note that there is no visible broadening in the reflection shown in Figure 14.12 as well as all others recorded. Accumulation times for one reflection using an area detector which covers 6°x6° were approximately 15 minutes for λ=0.71 Å and more than an hour for λ=0.33 Å due to the lower flux at this wavelength. The shadowing of the two pillars can be overcome by an anvil rotor, see section 12.4 for a description, but the azimuthal limitations will remain as long as Bridgman-type anvils are used. A full structural study will therefore require several runs with crystals in different orientations which can lead to a considerable amount of beam time.

The probably most challenging single-crystal neutron diffraction studies have been carried out in diamond anvil cells, on solid deuterium. Beside the extremely small sample volume the in-situ growth of a perfect single crystal in an adequate orientation is an additional experimental difficulty.

Glazkov et al. [180] presented diffraction data at 300 K up to 31 GPa obtained in the DAC shown in Figure 7.9. They were collected at the DISK low-background multidetector diffractometer of the IR-8 research reactor (thermal power: 6 MW, now decommissioned) of the Kurchatov Institute using a wave-

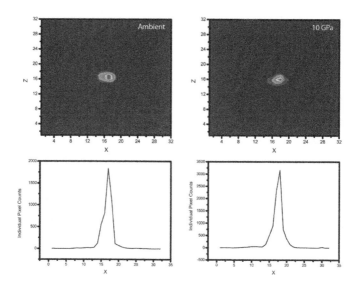

FIGURE 14.12

Single-crystal diffraction with a PE-press: two-dimensional images of the (400) reflection of a 3 mm^3 crystal of squaric acid ($H_2C_4O_4$) at ambient and 10 GPa. The corresponding d-spacings are 1.535 Å and 1.456 Å. Reprinted with permission from Ref. [179].

length of $\lambda=1.9$ Å. Deuterium solidifies at 5.5 GPa in a hcp lattice where the D_2 molecules are orientationally (and dynamically) disordered. Optical and x-ray diffraction studies indicate no phase transition at 300 K to at least 100 GPa. The primary aim was therefore to study the equation-of-state and compare it to various predictions. Figure 14.13 shows diffraction patterns of the strongest Bragg peaks ((002) and (101)) up to 31 GPa; in total three reflections could be observed up to this pressure, on a single crystal of 10^{-2} mm^3 starting volume, using diamonds with a culet size of 0.7 mm. In a different loading with a larger single crystal of 0.1 mm^3, five other reflections (100), (102), (110), (103), and (200) could be observed up to 6.2 GPa. The data collection time for a single reflection was 3-12 hours (S. Besedin, priv. comm.).

A similar experiment was carried out more recently at the 6T2 diffractometer of the LLB by Goncharenko and Loubeyre [70] to higher pressures (38 GPa) using a DAC as shown in Figure 7.10 and a sample size very similar to Glazkov et al.'s experiments. The aim of this experiment was to search for possible molecular ordering across the the 'broken symmetry phase' which occurs under pressure at low temperatures, at 38 GPa at 44 K. This involves hence the detection of changes in the peak intensity and the search for new reflections. A key strategy in these experiments was the pre-characterization

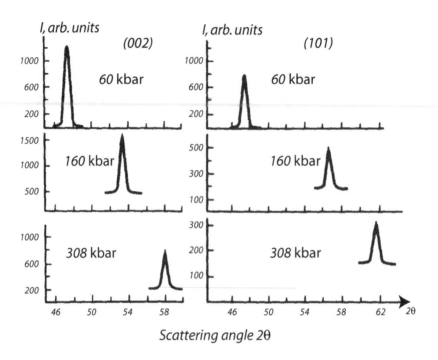

FIGURE 14.13
Single-crystal diffraction with a diamond anvil cell: (002) and (101) reflection
of solid deuterium D_2 at 300 K, measured at different pressures. Adapted from
Ref. [180].

using synchrotron x-ray diffraction which gave the orientation of the crystal
and which detected an additional satellite reflection which then could be studied further by neutron diffraction. In total 7 reflections were followed up to
38 GPa. As seen in Figure 7.10, the data are unable to detect a significant
change in intensity in any of the reflections, but show a reversible appearance
of a (1.07,0,0) superstructure satellite. The measurements were repeated in
2005 with a different single crystal up to 37 GPa. These unpublished data are
of better quality and indicate a small decrease of the (100) intensity by 1-2
e.s.d. and again the appearance of a satellite reflection when passing into the
broken symmetry phase (priv. comm. P. Loubeyre). In conclusion, both the
Russian and French experiments demonstrate that neutron diffraction in the
several-ten GPa range is possible using more or less standard diamond anvil
techniques. But they also show the limitations in terms of resolution, i.e. number of observable reflections, which would need shorter wavelengths (to access

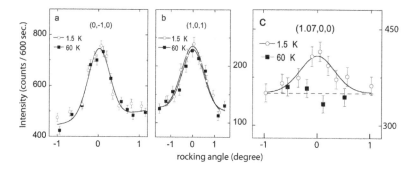

FIGURE 14.14

Single-crystal diffraction with a diamond anvil cell at low temperatures: Rocking scans of (0-10) and (101) reflections of solid deuterium D_2 at 38 GPa, in two different phases. Adapted from Refs. [70, 181].

larger parts of reciprocal space) and larger single crystals (to have sufficient intensity to detect them). The currently most realistic alternative appears to be the use of polycrystalline diamond (SD or NPD in the near future, see chapter 4.4) where it has been demonstrated that full structural determination is possible on mm^3 single crystals to 10 GPa (see above) and beyond [114]. The availability of gas loading methods (chapter 12.3) opens a realistic perspective to carry out such studies even in simple molecular systems, if high quality single crystal of this size can be grown under pressure.

14.2.3 Disordered systems

The study of disordered systems, i.e. liquids and glasses, is a prime topic in neutron scattering which exploits the ability to collect data to large Q-values and use isotopic substitution to extract partial structure factors. A general review on this topic can be found in Ref. [182].

The application of high pressure in such investigations has become more and more popular recently with the data analysis involving however various degrees of sophistication: (a) A simple observation of 'peak' position and width as function of P and T, with the aim to detect 'transitions' or structural relaxations. (B) The determination of partial structure factors by isotope substitution. This requires large Q coverage and careful normalization of the data with an accuracy in the per cent level. (C). The determination of partial structure factors in both the radial and angular degree of freedoms, i.e. determination of the spacial distribution functions, using in the simplest case a single data set, i.e. without use of isotopic substitution. This is only possible by introducing additional constrains, i.e. it involves a certain degree of modeling. Type (A) measurements can be relatively fast, but the amount of

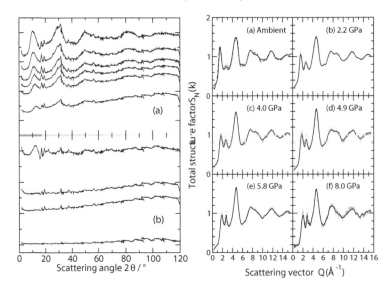

FIGURE 14.15

Diffraction patterns of GeO$_2$ glass under pressure, collected in angle-dispersive measurements at D4 (ILL). Left: Raw data at different pressures between 2.2 and 8 GPa (upper panel), and data of empty gasket compressed at different loads (lower panel). Right: Total structure factor at different pressures; verical bars indicate data, the lines are Fourier back-transforms of G(r). Reprinted with permission from Ref. [187].

structural information which can be extracted is very limited. Examples can be found in work on amorphous ice, see Refs. [183, 184]. Type (C) measurements give the most detailed structural information, but the results risk to be biased by the model. For complex systems this analysis should be rather considered a simulation. Typical examples where this method was successfully applied are measurements on liquid water and amorphous ice under pressure in the 0-7 GPa range [185, 186]. Whatever strategy is chosen, a key ingredient in all such measurements is a careful determination of the background and accurate correction for anvil occlusion and absorption. In the following we give two examples on GeO$_2$ which illustrate the principal difficulties encountered in such measurements.

In measurements at the D4 hot-source diffractometer at the ILL, Salmon et al. [187] studies germania (GeO$_2$) to 8.5 GPa. GeO$_2$ is a textbook example of a network-forming glass and has been studied under high pressure using numerous techniques. These measurements were carried out using a PE-press in an angle-dispersive mode, with the incident and diffracted beams passing through the TiZr gasket. The difficulties in these measurements are related to the fact that the amount of gasket material in the beam is about the same

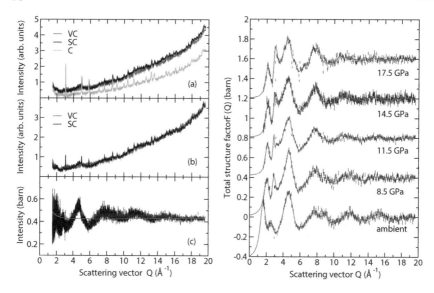

FIGURE 14.16

Diffraction patterns of GeO_2 glass under pressure, collected in time-of-flight measurements at PEARL (ISIS). Left: Measured intensities illustrating the raw data (a) and first steps in background correction ((b) and (c)). Labels correspond to measurements with the sample in the container (SC), with a piece of vanadium in the container (VC) and with the empty container (C). Right: Total structure factor at different pressures; vertical bars indicate data, the lines are spline fits to the data. After Salmon et al., unpublished.

as that of the sample, and that the geometry changes under pressure, see Figure 14.15 left panel. The signal of the gasket has hence to be measured in a separate run, at different levels of deformation which correspond to those with the sample under pressure.

The situation is quite different when such data are recorded at time-of-flight instruments, in transverse scattering. Here the influence of the TiZr gasket is negligible since it is not in the direct beam. Instead, the background is characterized by a steep rise at low d-spacings, i.e. high Q values which is related to secondary scattering of high-energy neutrons. Another non-trivial problem is the presence of Bragg edges from the anvils (see Figure 4.7 and Appendix B), which leads to a strongly distorted beam seen by the sample. For anvil materials with strong Bragg edges like diamond the effect is easily visible on the raw diffraction patterns and needs to be corrected. Figure 14.16 (left panel) illustrates the essential steps in background correction, after Salmon et al. It involves data collection on (i) the sample in its container (gasket), (ii) a vanadium sample in the container, and (iii) an empty container, each for several loads (pressures) to take care of the changing geometry and the slight pressure dependence of the Bragg edges. Note that the signal from the sample is hardly visible if the raw data are plotted over the full Q-range. Application of suitable interpolation schemes then leads to a data quality similar to what is obtained in angle-dispersive diffraction on D4. The advantage of this scattering geometry is that double-toroidal anvils can be used to reach pressures in the 20 GPa range (Figure 14.16, right panel). The same would be very difficult in angle-dispersive scattering due to the much less favorable sample/TiZr volume ratio. The disadvantage is the lack of data below Q$\approx$1.55 $\overset{\circ}{A}$ which contains often important features such as in GeO$_2$ the first sharp diffraction peak which occurs at 1.5 $\overset{\circ}{A}^{-1}$.

14.3   Inelastic scattering

In general one might distinguish between experiments where (a) discrete excitations are detected which are either coherent (such as phonons and magnons) or dispersionless (such as crystal field excitations), and (b) where S($\mathbf{q}$,ω) is spread out over a large ($\mathbf{q}$,ω) region with no obvious sharp features. Examples for the latter are scattering from spin fluctuations in paramagnets and weak ferromagnets, as well as phenomena often observed in intermediate valence and heavy fermion systems. Here we only give examples of (a) where a reasonable amount of well documented data exists.

14.3.1   Phonons

The aim here is to observe the effect of pressure on lattice vibrations which — in the ideal case — should be measured on single crystals. Measurements on powders provide only the (neutron weighted) phonon density of states (PDOS), but have the advantage to be much less time consuming and to rely on samples in a form which is much more easily available than single crystals. A common motivation of high pressure studies is the determination of Grüneisen parameters, in particular mode Grüneisen parameters $\gamma_{i,\mathbf{q}}$:

$$\gamma_{i,\mathbf{q}} = -\partial ln E_{i,\mathbf{q}}/\partial ln V \qquad (14.1)$$

where E$_{i,\mathbf{q}}$ is the energy of a phonon mode i and wave vector $\mathbf{q}$, and V the volume. Mode Grüneisen parameters are usually positive with magnitude around 1. They can be directly related to the thermal expansion: compounds with negative thermal expansion have necessarily negative Grüneisen parameters for some modes i and wave vectors $\mathbf{q}$. A considerable part of high pressure lattice dynamics studies is therefore motivated by understanding negative thermal expansion.

Figure 14.17 shows an example of a powder study on ice using a PE press at IN6 (ILL, Grenoble). Ice exhibits negative thermal expansion below 60 K

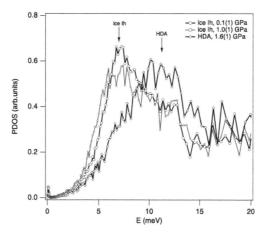

FIGURE 14.17
PDOS of ice to 1.6 GPa, obtained from measurements in a PE-cell at IN6 (ILL). Note the initial decrease of phonons below 7 meV under pressure. After Ref. [188].

and there are hence low-energy phonons which soften under pressure. The measurements were carried out at 80 K using cBN anvils and TiZr gaskets, with no pressure transmitting medium. These measurements exploit the large incoherent cross section of hydrogen which enables measurements on samples in the 10 mm³ range. At low energies the PDOS has the characteristic $\sim E^2$ dependence and peaks at 7 meV where the acoustic phonon branches flatten close to the zone boundary. Fig. 14.17 shows that under pressure to 1 GPa the entire PDOS below 7 meV softens till the sample has transformed fully to high density amorphous ice (HDA). Further compression in this phase leads then to a hardening of the phonons. These data can be compared to lattice dynamical models to identify which of the acoustic phonons are responsible for the observed effects and connect the phenomenon to the pressure-induced amorphisation.

Similar studies have been carried out on various negative thermal expansion materials, for example in ZrW_2O_8 [189] where a pressure of 0.2 GPa is enough to produce visible effects on the PDOS. These experiments were hence carried out in a gas cell.

Figure 14.18 shows an example of data obtained on C_{60} using a steel clamp cell with 6 mm inner and 12 mm outer diameter [190]. The experiments up to 0.9 GPa required a pressure transmitting medium, in this case Fluorinert, which contributes significantly to the measured signal. Figure 14.18 (left) illustrates the contributions from the pressure cell and the PTM which have to be subtracted. Note that the PTM shows a marked pressure dependence itself which needs to be determined in a separate measurement. At 300 K, C_{60}

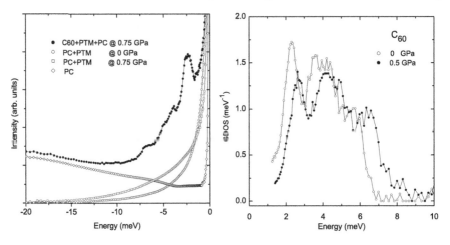

FIGURE 14.18
Phonon density of states (PDOS) measurement on C_{60} carried out with a steel clamp cell. Left: data illustrating the contributions of the sample, the pressure cell (PC) and the pressure transmitting medium (PTM). Negative energies indicate up-scattering. Right: PDOS of external modes at ambient and 0.5 GPa (300 K). Courtesy of H. Schober; partially redrawn form original plots in Ref. [190].

undergoes in the 0-0.5 GPa range a pressure-induced transition from a rotationally disordered to an ordered phase which might be visible in the PDOS, in particular in the libron bands which are around 2 meV. Ambient pressure data show indeed a softening of the librons as the transition is approached under cooling. The data (Figure 14.18) exhibit in contrast a general hardening of the PDOS with a Grüneisen parameter which is approximately constant. Data were also collected at moderately low temperatures down to 180 K.

The most detailed information on phonon dispersion under pressure needs single crystals and measurements on triple-axis spectrometers. For high pressure neutron scattering it means the compression of mm-size samples under sufficiently hydrostatic conditions to prevent breaking. Classical examples of high pressure experiments in the 0-0.5 GPa range using gas/liquid cells are the work on solid neon to 0.4 GPa [191], on solid deuterium to 0.45 GPa [192], on RbI [193] and KBr [194] to 0.4 GPa, and ice to 0.5 GPa [195]. A clamped CuBe cell served for measurements on SmS to 0.6 GPa [196], and a McWhan-type cell for experiments on black phosphorus to 2.5 GPa [197]. The work on graphite is one of the rare examples carried out in a sapphire cell [198].

Figure 14.19 illustrates fairly detailed measurements done on the phonon dispersion of semiconductors in the 0-10 GPa range using single crystals of 5-10 mm^3 volume compressed by a PE-load frame. Such reasonably robust samples can be compressed to multi-GPa pressures without damage if embedded in

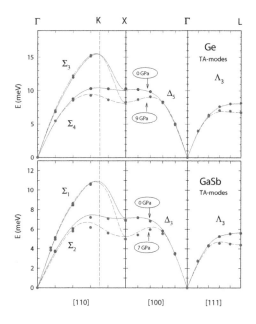

FIGURE 14.19
Phonon dispersion of transverse acoustic (TA) modes in cubic semiconductors Ge and GaSb under pressure, measured with a PE-cell. Reprinted with permission from Ref. [162].

lead using the method illustrated Figure 13.5, applying either TC or SD anvils with a profile shown in Figure D.2 (Appendix D). Data on Ge were collected at 1T1 at the LLB, the data on GaSb at IN8 (ILL), both are thermal triple-axis spectrometers. Typical accumulation times for one phonon scan were 1-3 hours, but strongly depending on the type of phonon. Similar to single-crystal diffraction studies using toroidal anvils, the proper alignment of the sample is crucial. The data show a pronounced phonon softening for transverse acoustic modes close to the zone boundary, characteristic for all tetrahedrally bonded semiconductors. This leads to the well known negative thermal expansion coefficient observed at low temperatures.

14.3.2 Magnons

Compared to phonon measurements, high pressure magnon dispersion data are very scarce, probably because of the experimental complexity involving in most cases both single crystals and low temperatures. In principle, such experiments allow the determination of magnon stiffness constants and hence the distance-dependence of the exchange interaction. A classical example carried out in a high pressure gas cell to 0.4 GPa is the work by Vettier and Yelon [199] on $FeCl_2$. This is a two-dimensional ferromagnet with large anisotropy which shows a weakly first-order phase transition at 0.2 GPa. A strong pressure-induced shift to higher energy is observed across this transition.

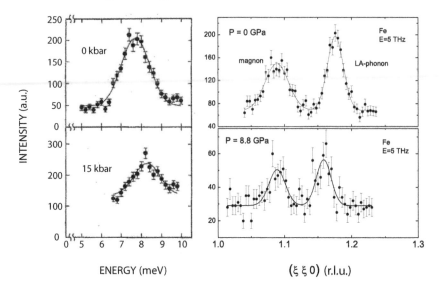

FIGURE 14.20

Magnons under pressure, observed in ferromagnetic terbium (left, T=90 K, ref. [200]) and iron (right, T=300 K, M. Braden et al., priv. communication). The data on Tb are constant-$\mathbf{Q}$ scans at $\mathbf{Q}$=(0,0,2.7), the data on Fe constant energy scans at E=5 THz=20.7 meV along [110]. Tb data redrawn from Ref. [200].

Figure 14.20 gives examples of the elemental ferromagnets terbium [200] and iron (M. Braden et al., unpublished), obtained with a McWhan-cell and a PE-cell, respectively. Terbium has a Curie temperature of 220 K which allows magnon dispersion measurements at moderately low temperatures of 90 K. In the measurements by Kawano et al. [200] a single crystal sapphire cell was used up to 0.4 GPa since it gave a low background, as well as Fluorinert as PTM. Data beyond this pressure were collected in a McWhan cell, Figure 14.20 shows some of the raw data. The entire dispersion along [110] was measured up to 1.5 GPa and fitted to a magnon dispersion relation (lines in Figure 14.21) which allowed the determination of the pressure dependence of the interplanar exchange and anisotropy parameters.

Magnon frequencies to considerably higher pressures could be determined using a PE-cell, on a 8 mm^3 single crystal of iron, embedded in lead as explained in Figure 13.5. The experimental difficulty is that magnon energies increase $\propto q^2$ which means that their measured intensities decrease rapidly with q. For low q values their energies in iron are close to those of the longi-

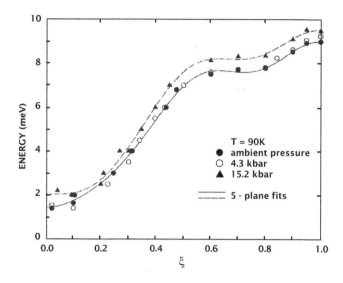

FIGURE 14.21
Magnon dispersion of terbium along [110], measured with a McWhan cell.
Reprinted with permission from Ref. [200].

tudinal acoustic (LA) mode and can therefore me recorded in the same scan.
Figure 14.20 demonstrates that the magnon frequencies in iron are rather
pressure insensitive compared to the LA-phonons which show their expected
pressure-induced hardening [201].

14.3.3 Crystal field excitations

Crystal field excitations in metals count among the first applications of high
pressure neutron scattering in the multi-GPa range. What is probed are the
transitions between the electronic levels split by the crystal electric field (CEF)
of the surrounding ligands. These splittings are in the 1-1000 meV range and
accessible to inelastic neutron scattering. The vast majority of cases concerns
rare earth elements where the CEF leads to a splitting of the (2J+1)-fold
degenerate ground state multiplet of the 4f-electrons. Inelastic neutron investigations of the CEF on rare earth compounds are rather favorable since
the cross section is proportional to J(J+1) (example: J=8 for Ho^{3+}), i.e. the
required amount of sample can be small.

As an example we cite here the classical work on PrSb by Vettier et al.
[202], one of the first experiments carried out in a McWhan cell, see Figure
14.22. PrSb is a model singlet-triplet system where the interplay of crystal-field
and exchange interaction can be studied. The well-established point-charge
model of the CEF would predict an increase of the energy levels proportional

to a^{-5} (a: lattice parameter). The measurements show that the contrary is the case, the exciton energies decrease along all direction with a Grüneisen parameter of $\approx$ -4.3. The softening suggests the occurrence of a phase transition to an ordered magnetic state at higher pressures, which indeed was found at 3 GPa [34].

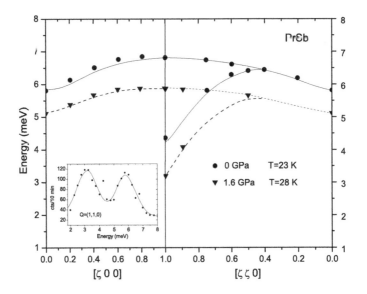

FIGURE 14.22
Pressure dependence of the crystal field excitations (excitons) in PrSb, measured with a McWhan cell. Adapted from Ref. [202].

Other examples concern the high-T_c superconductors which contain rare earth elements sandwiched between CuO_2 layers (example: $ErBa_2Cu_3O_x$). Since the CEF is sensitive to both the charge and the distance of the ligands, high pressure measurements on the CEF can give information on the pressure-induced charge transfer to the layers to explain the observed pressure dependence of T_c. Such experiments have indeed been carried out with a 1 GPa clamp cell, see Mesot et al. [203] for details.

Figure 14.23 illustrates an example on $NdAl_3$, measured in a clamp cell to 1 GPa, such as shown in Figure 5.7. The data were collected at the FOCUS time-of-flight spectrometer of the Swiss spallation source SINQ. $NdAl_3$ crystallizes in the hexagonal Ni_3Sn structure ($P6_3/mmc$) and remains paramagnetic over the whole temperature and pressure range considered. The CEF-Hamiltonian in this structure has four adjustable parameters which were determined by a fit through the data, see Figure 14.23. The pressure dependence of these parameters was then compared to predictions from the well known point-charge model as well as results from DFT calculations. Such high pressure measurements provide therefore a gauge for the reliability of rare earth CF-models under variable density.

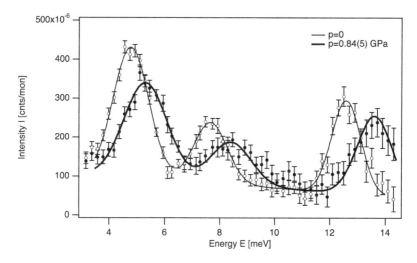

FIGURE 14.23
Pressure dependence of crystal-field excitations in NdAl$_3$ measured with a clamp cell. Reproduced with permission from Ref. [204].

14.4 Small angle neutron scattering (SANS)

SANS probes structures which are larger than typically 100 Å. The vast majority of investigations, including high pressure experiments, concerns soft condensed matter research. The study of low temperature magnetic phenomena is another important SANS application but high pressure measurements in this domain are rather rare.

An impressive example of the latter is the pressure- and field-induced detection of long-range helical order in MnSi at 1.8 GPa, measured at 1.5 K in a CuBe-clamp cell at the V4 small-angle instrument of the HMI [206], see Figure 14.24. The signal appears at Q=0.04 Å^{-1} which corresponds to a periodicity 150 Å. This example illustrates at the same time the drawback of CuBe-cells for SANS, i.e. the interference of the microstructure present in the pressure cell material (precipitations, grain boundaries etc.) with the signal of the sample, located in the q=0.01-0.2 Å^{-1} range.

As stated above, most applications of SANS concern soft condensed matter research where the relevant pressure scale is 0-1 GPa, and where temperatures remain close to 300 K. For this purpose, sapphire cells as discussed in chapter

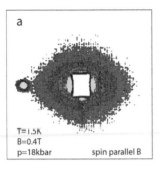
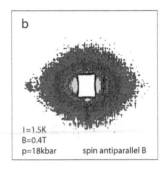

FIGURE 14.24

SANS pattern of MnSi recorded with a CuBe cell at 1.8 GPa and 1.5 K. The intensity around the center is due to the pressure cell. The feature of interest is the spot to the left-hand side which is due to the helical order as aligned in a magnetic field. Reproduced with permission from Ref. [205].

8.1 have become the key tool, for at least two reasons: (a) the very small background, and (b) the possibility for complementary measurements such as light scattering.

Figure 14.25 shows a typical example of a high pressure SANS study on a solution of a polymer (poly(ethylene-*co*-1-buten), PEB) in pentane [207]. The experiments were carried out with a sapphire-window cell at the NG-3 SANS spectrometer at NIST (Gaithersburg). A main objective of such experiments is the determination of correlation lengths ζ and the radius of gyration R_g which characterize the inter and intra-chain length scales, respectively. The two quantities are extracted from a fit to the measured total and single scattering intensities $S_t(q)$ and $S_s(q)$. The two quantities can be obtained from two data sets with adequate deuteration. The relevant formula are the Ornstein-Zernike and Debye equations:

$$S_t(q) = \frac{S(0)}{1 + q^2\zeta^2} \tag{14.2}$$

$$S_s(q) = \frac{2}{q^4 R_g^4}[exp(-q^2 R_g^2) + q^2 R_g^2 - 1] \tag{14.3}$$

In the example of Figure 14.25, ζ is observed to increase by a factor 5 whereas R_g is pressure-independent.

Another typical example is the study of surfactants under pressure, again investigated using sapphire window cells [73]. In solutions these macromolecules form micelles which organize themselves to form mesophases, i.e. partially ordered large-scale structures which can be characterized by SANS. Figure 14.26 shows high pressure data on an important surfactant, $C_{12}H_{25}OSO_3^-Na^+$ (sodium dodecyl sulfate, SDS), dispersed in water. Above

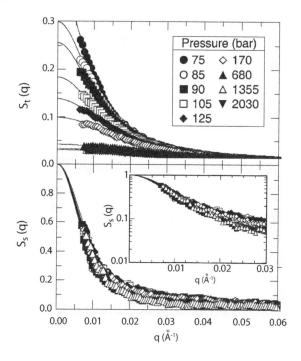

FIGURE 14.25

High pressure SANS patterns of a poly(ethylene-*co*-1-buten) (PEB) - pentane solution, carried out with a sapphire-window cell. Total chain ($S_t(q)$, top) and single chain ($S_s(q)$, bottom) scattering in a 5wt% PEB - d_{12} pentane solution, at 130 °C. Adapted from Ref. [207].

20 °C and up to a concentration of 35wt%, SDS forms spherical micelles of a diameter of 36 Å. Figure 14.26 (left) shows that for the 30wt% solution the correlation peak shifts under pressure to smaller Q values implying larger micelle sizes, probably due to pressure-induced elongation. In 50wt% solutions, the micelles order at ambient conditions in a hexagonal lattice with a = 51.5 Å. Under pressure (Figure 14.26, right panel) a phase transition occurs above 0.1 GPa to a possibly fully crystalline phase.

14.5   Quasi-elastic neutron scattering (QENS)

QENS exploits the incoherent cross section of certain elements, in particular hydrogen, for the study of the stochastic motion of atoms and molecules. For a review of this technique see Refs. [208, 209]. Typical examples are the diffusion

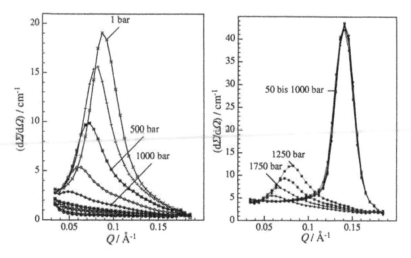

FIGURE 14.26

High pressure SANS patterns of surfactant ($C_{12}H_{25}OSO_3^-Na^+$ (SDS)), carried out with a sapphire-window cell. Left: Scattering from a 30wt% solution. Right: Scattering from a 50wt% solution. Reproduced and adapted with permission from Ref. [73].

of hydrogen in metals, or the motion of polymers. The most critical parameter in this technique is the instrumental resolution which defines the accessible timescales. Typical diffusive motion of small molecules are in the 1-100 ps range which requires energy resolutions in the meV to μeV range. The diffusion in polymers is much slower, in the μs to ns range, and therefore requires very high resolutions, in the order of 10^{-6}-10^{-3} meV. Since an increase of resolution always translates into a reduction of intensity, high pressure QENS studies become increasingly difficult as the timescale of the investigated motion increases.

Figure 14.27 illustrates a typical high pressure QENS measurement on BPTI[2], a small protein of 58 amino acid residues [27, 210]. The measurements were carried out in a CuBe cell very similar to the one shown in Figure 5.2. Diffusion coefficients and characteristic time scales of diffusion can be determined from the measured $S(\mathbf{Q},\omega)$ (lower panel) through its Q-dependence and its Fourier transform, respectively [211, 212]. The experimental difficulty is the separation of the contributions from the sample (BPTI), the pressure cell, and the buffer liquid which is added to avoid aggregation. It is evident that separate measurements (as shown in the upper left panel) need to be carried out at each pressure to derive meaningful $S(\mathbf{Q},\omega)$ values, particularly since the pressure dependence in the 0-0.6 GPa range is not immediately evident.

---

[2]Bovine pancreatic trypsin inhibitor.

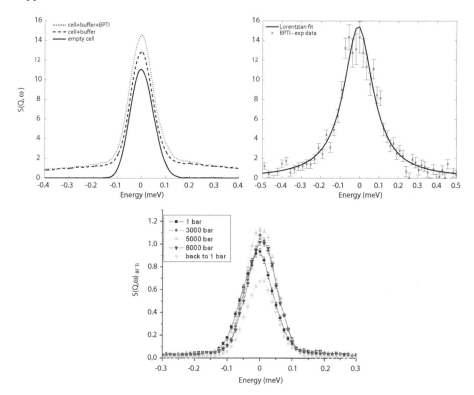

FIGURE 14.27
Typical QENS spectra obtained with a CuBe cell illustrating the various contributions (upper left) to the sample signal (upper right). The pressure dependence is shown in the lower panel. Reproduced with permission from Ref. [27, 210].

Another example is illustrated in Figure 14.28 showing data on H_2O, obtained at 400 K either in an aluminum alloy cell (upper row) and a PE-cell (lower row), both measured at IN6 (ILL). The Al-cell is identical to the one shown in Figure 5.2 and has a cylindrical Al-spacer inside to reduce the multiple-scattering. The sample volumes are 150 mm^3 and 40 mm^3, for the Al- and PE-cell, respectively. Here the timescales of diffusion are such that the width of the QENS signal (essentially a Lorentzian) can be easily separated from the signal of the cell which produces an elastic line with much smaller width. The strong pressure dependence of the width is clearly visible in this system.

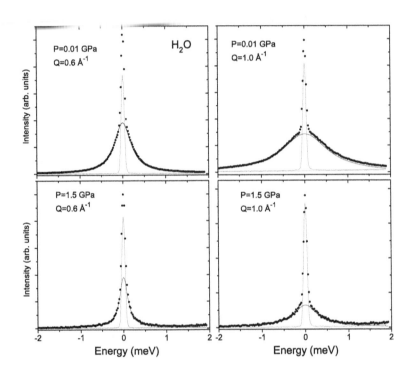

FIGURE 14.28

QENS pattern on liquid water under pressure. Top row: collected in an Al-alloy cell. Bottom row: collected in a PE cell using toroidal anvils. The measured (raw) data are marked by black dots, the measured gasket/cell contribution by dashed lines, and the background by dotted lines. The QENS signal from the sample is shown as a solid line. After Bove et al., to be published.

Appendix A

The Von Mises plasticity criterion

One of the principal problems of mechanical engineering is to predict the onset of plasticity for an object of arbitrary geometry subjected to an arbitrary load, knowing only one available quantity, the yield stress σ_Y obtained from traction experiments.

It is clear that plasticity is somehow related to shear and cannot be induced by pure hydrostatic pressure. This is easy to see, since a material subjected to pure hydrostatic pressure can be compressed to any volume and will return to its initial shape after pressure release, without plastic deformation, provided that it does not show a phase transition. For this reason the *deviatoric stress tensor* $\mathbf{s}=(s_{ij})$ is defined by:

$$\mathbf{s} = \boldsymbol{\sigma} - \frac{1}{3}Tr[\boldsymbol{\sigma}]\mathbf{1} \tag{A.1}$$

where $\boldsymbol{\sigma}=(\sigma_{ij})$ is the stress tensor, $\mathrm{Tr}[\boldsymbol{\sigma}]$ is the trace of $\boldsymbol{\sigma}$ (i.e. the sum over its diagonal elements) and $\mathbf{1}$ is the identity matrix. Since the second term is by definition the hydrostatic component of $\boldsymbol{\sigma}$, $\mathbf{s}$ is the stress tensor where the hydrostatic component has been subtracted. The *equivalent tensile stress* or *von Mises stress* σ_{VM} is then defined as:

$$\sigma_{VM} = \sqrt{\frac{3}{2}\sum_{i,j}(s_{ij})^2} \tag{A.2}$$

Note that the expression under the square root is to within a factor $3/2$ the second invariant of $\mathbf{s}$, i.e. σ_{VM} is independent of the choice of the particular coordinate system. For a given object submitted to a given stress field σ_{ij}, values of von Mises stresses σ_{VM} can be readily calculated and then visualized by plotting "iso-Mises lines." For an arbitrary geometry such calculations can only be done numerically by applying so called Finite Element (FE) methods. Iso-Miso lines are a standard output of all FE-softwares.

To grasp the physical meaning of the definition A.2, let us choose a coordinate system in which $\boldsymbol{\sigma}$ is diagonal with coefficients σ_1, σ_2, σ_3, the *principal stresses*. It is always possible to define such a reference system since $\boldsymbol{\sigma}$ (and $\mathbf{s}$) are symmetric matrices. With this choice, by inserting $\mathbf{s}$ into equation A.2 we obtain after some algebra:

$$\sigma_{VM} = \sqrt{((\sigma_1 - \sigma_2)^2 + (\sigma_1 - \sigma_3)^2 + (\sigma_2 - \sigma_3)^2)/2} \tag{A.3}$$

which for a simple traction along z ($\sigma_1 = \sigma_2 = 0$) becomes:

$$\sigma_{VM} = \sigma_3 \tag{A.4}$$

In other words, in the case of simple traction, such as carried out on test rods to determine the yield stress σ_Y, the von Mises stress is identical to the applied stress, i.e. the force on the test rod divided by its cross section. Plasticity occurs hence when $\sigma_{VM} = \sigma_Y$. The *von Mises plasticity criterion* generalizes this to any stress field, i.e. it is claimed that plasticity occurs whenever

$$\sigma_Y = \sqrt{\frac{3}{2} \sum_{i,j} (s_{ij})^2} \tag{A.5}$$

Note that in a cylindrical coordinate system the principal stresses are σ_r, σ_θ, and σ_z, and equation A.3 is identical to equation 1.18.

Appendix B

Neutron attenuation

When a neutron beam crosses a material it loses its intensity, an affect which we call attenuation. Attenuation occurs through three different processes: (a) pure absorption, i.e. neutron capture, (b) coherent scattering, and (c) incoherent scattering. The corresponding cross sections are usually labeled σ_{abs}, σ_{coh}, and σ_{inc}. These cross sections are wavelength (energy) dependent, and the standard tables give them for a wavelength of 1.7898 Å. For medium to large wavelengths λ one finds usually $\sigma_{abs} \propto \lambda$. For short wavelengths (high energies) most materials have so-called "resonances", i.e. sharp absorption bands at well defined energies where σ_{abs} increases by several orders of magnitude. As for σ_{coh} and σ_{inc}, the simplest model is a ballistic approach where no wavelength dependence is assumed. This is fairly correct for σ_{inc}, but not for σ_{coh}, since for wavelengths in the order of the lattice spacing coherent diffraction occurs, a process which scatters neutrons out of the incident beam. To be specific, the transmission (=fraction of transmitted to incident intensity) of a neutron beam crossing a homogeneous and isotropic material in form of a slab of thickness x is given by [213]:

$$T = \exp[-x\{N\sigma_{coh} + n(\sigma_{inc} + \sigma_{abs})\}] \tag{B.1}$$

where N is the number of unit cells per volume and n the number of atoms per volume. The wavelength dependence of the coherent elastic cross section is given by [213]:

$$\sigma_{coh}(\lambda) = \sum_{\tau \, for \, 2d_\tau > \lambda} \frac{\lambda^2 z_\tau}{2V_{cell}} e^{-2W} |F(\tau)|^2 d_\tau \tag{B.2}$$

where V_{cell} is the unit cell volume, e^{-2W} the Debye Waller factor, and $F(\tau)$, z_τ and d_τ the structure factor, the multiplicity and the d-spacing at the reciprocal lattice vector τ, respectively. Note that for a given wavelength, the sum is taken over reciprocal lattice vectors τ constrained by $2d_\tau > \lambda$. The reason is that according to Bragg's law $\lambda = 2d_\tau sin(\theta)$ only these vectors τ can contribute to elastic scattering and hence to the attenuation of the neutron beam. Therefore, if τ_{max} is the largest reciprocal lattice vector, $\sigma_{coh}(\lambda) = 0$ for wavelengths $\lambda > 2d_{\tau_{max}}$.

To illustrate the general behavior of $\sigma_{coh}(\lambda)$ as a function of wavelength let us take the example of diamond, see Figure 4.7 in this context. The lattice parameter of diamond is $a = 3.57$ Å and the largest reciprocal lattice vector corresponds to the (111) type reflections with $d_{(111)} = 2.06$ Å. Hence, the coherent elastic cross section is zero for wavelengths larger than $2 \cdot 2.06$ Å $= 4.12$ Å. Beyond this wavelength attenuation can only be caused by pure absorption (neutron capture) or incoherent scattering given by σ_{abs} and σ_{inc}, respectively. When the wavelength is slightly below 4.12 Å, the sum in B.2 extends over all 8 equivalent (111) reflections, and accordingly σ_{coh} has a finite value. In other words the total absorption increases sharply. This is called a *Bragg edge*, see Figure 4.7. When the wavelength is further reduced the coherent cross section decreases smoothly $\propto \lambda^2$ according to equation B.2 until the condition $\lambda > 2d_\tau$ is also satisfied for the $\tau=(220)$ reflections for which $2d_{(220)} = 2.52$ Å. At this wavelength σ_{coh} again increases sharply giving rise to another Bragg edge. From equations B.1 and B.2 it is clear that Bragg edges are most pronounced for materials with low absorption and incoherent cross sections and strong scatterers, i.e. materials with large coherent cross section and structures of high symmetry which leads to large structure factors $F(\tau)$. Diamond is a typical example. Bragg edges in cemented tungsten carbide are much less pronounced (Figure 4.7).

For a simple estimation of the attenuation of a thermal neutron beam of wavelength λ, i.e. ignoring diffraction effects as described above, one can take the values for σ_{coh}, σ_{inc}, and σ_{abs} tabulated in Ref. [214] and Appendix G for $\lambda_0=1.798$ Å and calculate the total cross section:

$$\sigma_{tot}(\lambda) = \sigma_{coh} + \sigma_{inc} + (\lambda/\lambda_0)\sigma_{abs} \tag{B.3}$$

i.e. only the term which contains the absorption cross section is wavelength dependent. This formula can be easily generalized for materials containing not one but several elements.

Then the $1/e$ attenuation length l is given by:

$$l = \frac{1}{n\sigma_{tot}(\lambda)} \tag{B.4}$$

where n is again the number of atoms per volume, i.e. the number density. This is how various absorption calculators available on the internet (the NIST website for example) derive attenuation lengths.

Appendix C

The frettage of two cylinders

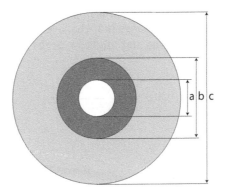

FIGURE C.1
Compound cylinder made of two different materials. The inner cylinder ("liner") is under radial compression of the outer cylinder ("jacket").

Suppose a geometry as illustrated in Figure C.1 where an inner cylinder (liner) with inner and outer diameter a and b, respectively, is radially supported by an outer cylinder (jacket) of inner and outer diameter b and c. Before assembling the two parts, the outer diameter of the liner is slightly *larger* than the inner diameter of the jacket, which means that the liner is under *compression* of the jacket once the two parts are assembled. The assembly is done by either shrink fitting the two parts (in which case the interface has a slightly conical angle of typically 1-2°) or by heating the jacket to expand its diameter sufficiently enabling the liner to fit into it. Such radially supported items are called frettages (from French: frettage; German: Armierung).

Given a desired radial pressure P at the interface, the *diametral interference* δ of the two cylinders (i.e. the difference in initial outer diameter of the liner and the initial inner diameter of the jacket) is then given by [3]:

$$\delta = \frac{bP}{E_o}[\frac{K_o^2+1}{K_o^2-1}+\nu_o] + \frac{bP}{E_i}[\frac{K_i^2+1}{K_i^2-1}-\nu_i] \qquad (C.1)$$

where

$K_i = b/a$ and $K_o = c/b$ are the wall ratios of the inner and outer cylinder, and E_o/ν_o and E_i/ν_i are the respective Young moduli / Poisson's ratios.

If the inner cylinder is a die, i.e. a solid rod or disc ($a \to 0$), equation C.1 becomes:

$$\delta = \frac{bP}{E_o}[\frac{K_o^2 + 1}{K_o^2 - 1} + \nu_o] + \frac{bP}{E_i}[1 - \nu_i]$$ (C.2)

A frequently encountered case is a steel-enforced backing plate made of a hard material, in general tungsten carbide, see for example Figure 10.2. For tungsten carbide $E \approx 600$ GPa which is about three times larger than that of steel. With K_o typically 3 and $\nu \approx 0.3$, the second term in C.2 is about six times smaller than the first one and might be neglected in first approximation to give:

$$\delta = \frac{bP}{E_o}[\frac{K_o^2 + 1}{K_o^2 - 1} + \nu_o]$$ (C.3)

Appendix D

Toroidal anvil profiles

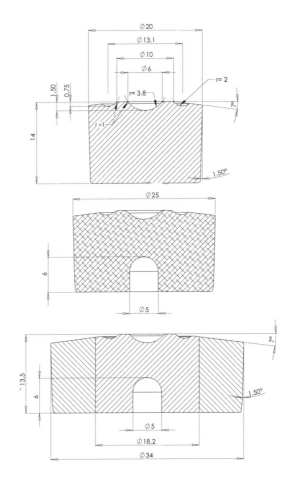

FIGURE D.1

Single-toroidal anvil profile used in the majority of current high pressure neutron scattering experiments on powders ("Los Alamos profile"). The anvil materials are cBN (upper), TC (middle) and sintered diamond/TC (lower). The dimensions of the profile are identical for all three cases.

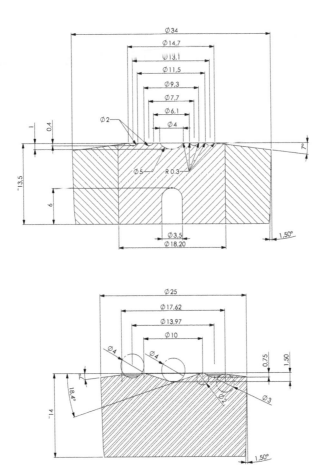

FIGURE D.2
Upper: Double-toroidal anvil profile used for high pressure experiments to 30 GPa. Materials: Sintered diamond/TC. Lower: Single-toroidal anvil profile used for single crystal inelastic neutron scattering. The anvil material is tungsten carbide.

Appendix E

Melting curves

- ## Helium

TABLE E.1

Melting data of ⁴He after McCarty [215]. 1 bar = 0.1 MPa

T_m(K)	P_m(bar)	liquid density (mole/l)
2.0	37.74	46.18
2.5	57.10	48.67
3.0	79.96	50.97
3.5	105.21	52.94
4.0	132.22	54.63
5.0	191.17	57.54
6.0	258.15	60.25
7.0	332.82	62.85
8.0	413.68	65.30
9.0	500.35	67.63
10	592.57	69.85
12	792.57	73.91
14	1012.06	77.37
16	1249.71	80.43
18	1504.48	83.34
20	1775.51	86.07
22	2062.01	88.67
24	2363.39	91.17
26	2679.04	93.60
28	3008.48	95.98
30	3351.26	98.34

For the pressure range up to 10 GPa, the melting curve of ⁴He can be expressed by a Simon equation [159]:

$$P(GPa) = 0.001767 \, T^{1.644} - 0.003 \qquad (E.1)$$

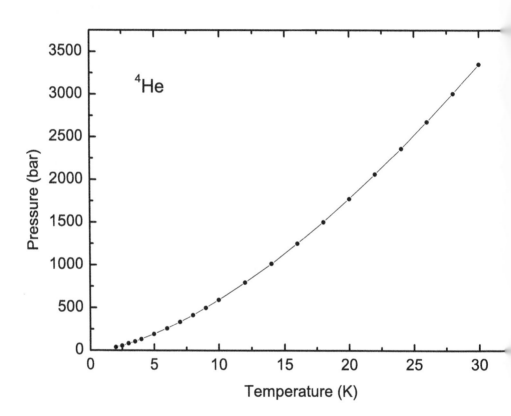

FIGURE E.1
Melting curve of ^{4}He after McCarty [215], see Table E.1 for numerical values.

• 5:1 Iso/n-pentane

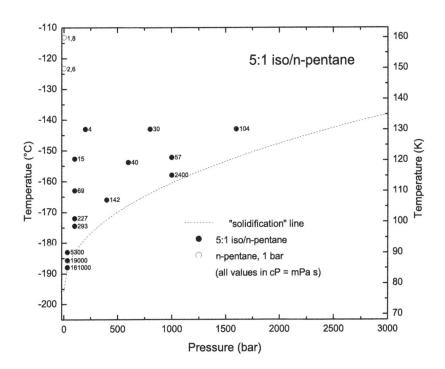

FIGURE E.2

Viscosity of 5:1 iso/n-pentane as a function of temperature and pressure. The line indicates a viscosity of approximately 10-100 Pa s where a capillary of 0.2 mm inner diameter and 1 m length appears to be blocked. Adapted from Ref. [216].

• 4:1 Methanol-Ethanol

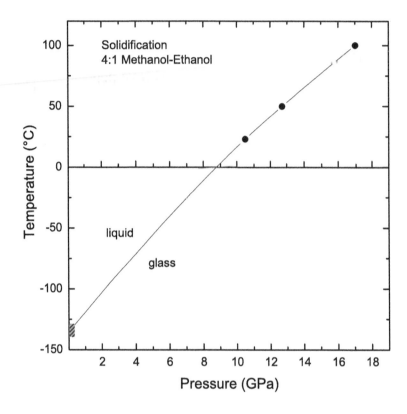

FIGURE E.3

Solidification of 4:1 methanol-ethanol as a function of temperature and pressure, derived from ruby fluorescence measurement in a diamond anvil cell. The dashed area indicates the temperature range where the mixture solidifies at ambient pressure. Adapted from Ref. [217].

Appendix F

Useful relations and conversion factors

Relation between wavelength λ and energy E of a neutron:

$$E[meV] = \frac{80.80}{\lambda^2[\mathring{A}]} \quad \text{(F.1)}$$

Relation between wavelength λ and speed v of a neutron:

$$v[m/s] = \frac{3956}{\lambda[\mathring{A}]} \quad \text{(F.2)}$$

Relation between momentum transfer Q, d-spacing and scattering angle 2θ (Bragg's law):

$$Q = \frac{2\pi}{d} = \frac{4\pi sin(\theta)}{\lambda} \quad \text{(F.3)}$$

TABLE F.1
Units of pressure and conversion factors

	bar	Pa	kg/cm^2	atm	lb/in^2	mm Hg
1 bar	1	10^5	1.01971	0.98692	14.5038	750.062
1 Pa	10^{-5}	1	1.01971×10^{-5}	0.98692×10^{-5}	14.5038×10^{-5}	750.062×10^{-5}
1 kg/cm^2	0.98067	0.98692×10^5	1	0.96785	14.2234	735.55
1 atm	1.01325	1.01325×10^5	1.03322	1	14.6960	760
1 lb/in^2	0.068947	6.894×10^3	0.070306	0.068045	1	51.715
1 mm Hg	1.33322×10^{-3}	1.33322×10^3	0.0013595	0.013158	0.019337	1

1 Pa = 1 N/m^2
1 bar = 10^6 dynes/cm^2

Appendix G

Neutron scattering lengths and cross sections

TABLE G.1: Neutron scattering lengths and cross sections (in barn) of the elements and corresponding isotopes [214].

Isotope	conc (%)	b_{coh} (fm)	b_{inc} (fm)	σ_{coh}	σ_{inc}	σ_{tot}	σ_{abs}
H	—	-3.7390	—	1.7568	80.26	82.02	0.3326
^{1}H	99.985	-3.7406	25.274	1.7583	80.27	82.03	0.3326
^{2}H	0.015	6.671	4.04	5.592	2.05	7.64	0.000519
^{3}H	(12.32 a)	4.792	-1.04	2.89	0.14	3.03	0
He	—	—	—	1.34	0	1.34	0.00747
^{3}He	0.00014	5.74-1.483i	-2.5+2.568i	4.42	1.6	6	5333.(7.)
^{4}He	99.99986	3.26	0	1.34	0	1.34	0
Li	—	-1.90	—	0.454	0.92	1.37	70.5
^{6}Li	7.5	2.00-0.261i	-1.89+0.26i	0.51	0.46	0.97	940.(4.)
^{7}Li	92.5	-2.22	-2.49	0.619	0.78	1.4	0.0454
Be	100	7.79	0.12	7.63	0.0018	7.63	0.0076
B	—	5.30-0.213i	—	3.54	1.7	5.24	767.(8.)
^{10}B	20	-0.1-1.066i	-4.7+1.231i	0.144	3	3.1	3835.(9.)
^{11}B	80	6.65	-1.3	5.56	0.21	5.77	0.0055
C	—	6.6460	—	5.551	0.001	5.551	0.0035
^{12}C	98.9	6.6511	0	5.559	0	5.559	0.00353
^{13}C	1.1	6.19	-0.52	4.81	0.034	4.84	0.00137
N	—	9.36	—	11.01	0.5	11.51	1.9
^{14}N	99.63	9.37	2.0	11.03	0.5	11.53	1.91
^{15}N	0.37	6.44	-0.02	5.21	0.00005	5.21	0.000024
O	—	5.803	—	4.232	0.0008	4.232	0.00019
^{16}O	99.762	5.803	0	4.232	0	4.232	0.0001
^{17}O	0.038	5.78	0.18	4.2	0.004	4.2	0.236
^{18}O	0.2	5.84	0	4.29	0	4.29	0.00016
F	100	5.654	-0.082	4.017	0.0008	4.018	0.0096
Ne	—	4.566	—	2.62	0.008	2.628	0.039
^{20}Ne	90.51	4.631	0	2.695	0	2.695	0.036
^{21}Ne	0.27	6.66	(+/-)0.6	5.6	0.05	5.7	0.67
^{22}Ne	9.22	3.87	0	1.88	0	1.88	0.046
Na	100	3.63	3.59	1.66	1.62	3.28	0.53
Mg	—	5.375	—	3.631	0.08	3.71	0.063
^{24}Mg	78.99	5.66	0	4.03	0	4.03	0.05
^{25}Mg	10	3.62	1.48	1.65	0.28	1.93	0.19
^{26}Mg	11.01	4.89	0	3	0	3	0.0382
Al	100	3.449	0.256	1.495	0.0082	1.503	0.231
Si	—	4.1491	—	2.163	0.004	2.167	0.171
^{28}Si	92.23	4.107	0	2.12	0	2.12	0.177
^{29}Si	4.67	4.70	0.09	2.78	0.001	2.78	0.101
^{30}Si	3.1	4.58	0	2.64	0	2.64	0.107
P	100	5.13	0.2	3.307	0.005	3.312	0.172
S	—	2.847	—	1.0186	0.007	1.026	0.53
^{32}S	95.02	2.804	0	0.988	0	0.988	0.54
^{33}S	0.75	4.74	1.5	2.8	0.3	3.1	0.54
^{34}S	4.21	3.48	0	1.52	0	1.52	0.227
^{36}S	0.02	3.(1.)	0	1.1	0	1.1	0.15
Cl	—	9.5770	—	11.5257	5.3	16.8	33.5
^{35}Cl	75.77	11.65	6.1	17.06	4.7	21.8	44.1
^{37}Cl	24.23	3.08	0.1	1.19	0.001	1.19	0.433
Ar	—	1.909	—	0.458	0.225	0.683	0.675
^{36}Ar	0.337	24.90	0	77.9	0	77.9	5.2
^{38}Ar	0.063	3.5	0	1.5(3.1)	0	1.5(3.1)	0.8
^{40}Ar	99.6	1.830	0	0.421	0	0.421	0.66

(note: 1 fm = 10^{-15} m, 1 barn = 10^{-24} cm^2)

TABLE G.1: (continued)

Isotope	conc (%)	b_{coh} (fm)	b_{inc} (fm)	σ_{coh}	σ_{inc}	σ_{tot}	σ_{abs}
K	—	3.67	—	1.69	0.27	1.96	2.1
^{39}K	93.258	3.74	1.4	1.76	0.25	2.01	2.1
^{40}K	0.012	3.(1.)	—	1.1	0.5	1.6	35.(8.)
^{41}K	6.73	2.69	1.5	0.91	0.3	1.2	1.46
Ca	—	4.70	—	2.78	0.05	2.83	0.43
^{40}Ca	96.941	4.80	0	2.9	0	2.9	0.41
^{42}Ca	0.647	3.36	0	1.42	0	1.42	0.68
^{43}Ca	0.135	-1.56	—	0.31	0.5	0.8	6.2
^{44}Ca	2.086	1.42	0	0.25	0	0.25	0.88
^{46}Ca	0.004	3.6	0	1.6	0	1.6	0.74
^{48}Ca	0.187	0.39	0	0.019	0	0.019	1.09
Sc	100	12.29	-6.0	19	4.5	23.5	27.5
Ti	—	-3.438	—	1.485	2.87	4.35	6.09
^{46}Ti	8.2	4.93	0	3.05	0	3.05	0.59
^{47}Ti	7.4	3.63	-3.5	1.66	1.5	3.2	1.7
^{48}Ti	73.8	-6.08	0	4.65	0	4.65	7.84
^{49}Ti	5.4	1.04	5.1	0.14	3.3	3.4	2.2
^{50}Ti	5.2	6.18	0	4.8	0	4.8	0.179
V	—	-0.3824	—	0.0184	5.08	5.1	5.08
^{50}V	0.25	7.6	—	7.3(1.1)	0.5	7.8(1.0)	60.(40.)
^{51}V	99.75	-0.402	6.35	0.0203	5.07	5.09	4.9
Cr	—	3.635	—	1.66	1.83	3.49	3.05
^{50}Cr	4.35	-4.50	0	2.54	0	2.54	15.8
^{52}Cr	83.79	4.920	0	3.042	0	3.042	0.76
^{53}Cr	9.5	-4.20	6.87	2.22	5.93	8.15	18.1(1.5)
^{54}Cr	2.36	4.55	0	2.6	0	2.6	0.36
Mn	100	-3.73	1.79	1.75	0.4	2.15	13.3
Fe	—	9.45	—	11.22	0.4	11.62	2.56
^{54}Fe	5.8	4.2	0	2.2	0	2.2	2.25
^{56}Fe	91.7	9.94	0	12.42	0	12.42	2.59
^{57}Fe	2.2	2.3	—	0.66	0.3	1	2.48
^{58}Fe	0.3	15.(7.)	0	28	0	28.(26.)	1.28
Co	100	2.49	-6.2	0.779	4.8	5.6	37.18
Ni	—	10.3	—	13.3	5.2	18.5	4.49
^{58}Ni	68.27	14.4	0	26.1	0	26.1	4.6
^{60}Ni	26.1	2.8	0	0.99	0	0.99	2.9
^{61}Ni	1.13	7.60	(+/-)3.9	7.26	1.9	9.2	2.5
^{62}Ni	3.59	-8.7	0	9.5	0	9.5	14.5
^{64}Ni	0.91	-0.37	0	0.017	0	0.017	1.52
Cu	—	7.718	—	7.485	0.55	8.03	3.78
^{63}Cu	69.17	6.43	0.22	5.2	0.006	5.2	4.5
^{65}Cu	30.83	10.61	1.79	14.1	0.4	14.5	2.17
Zn	—	5.680	—	4.054	0.077	4.131	1.11
^{64}Zn	48.6	5.22	0	3.42	0	3.42	0.93
^{66}Zn	27.9	5.97	0	4.48	0	4.48	0.62
^{67}Zn	4.1	7.56	-1.50	7.18	0.28	7.46	6.8
^{68}Zn	18.8	6.03	0	4.57	0	4.57	1.1
^{70}Zn	0.6	6.(1.)	0	4.5	0	4.5(1.5)	0.092
Ga	—	7.288	—	6.675	0.16	6.83	2.75
^{69}Ga	60.1	7.88	-0.85	7.8	0.091	7.89	2.18
^{71}Ga	39.9	6.40	-0.82	5.15	0.084	5.23	3.61
Ge	—	8.185	—	8.42	0.18	8.6	2.2
^{70}Ge	20.5	10.0	0	12.6	0	12.6	3
^{72}Ge	27.4	8.51	0	9.1	0	9.1	0.8
^{73}Ge	7.8	5.02	3.4	3.17	1.5	4.7	15.1
^{74}Ge	36.5	7.58	0	7.2	0	7.2	0.4
^{76}Ge	7.8	8.2	0	8.(3.)	0	8.(3.)	0.16
As	100	6.58	-0.69	5.44	0.06	5.5	4.5
Se	—	7.970	—	7.98	0.32	8.3	11.7
^{74}Se	0.9	0.8	0	0.1	0	0.1	51.8(1.2)
^{76}Se	9	12.2	0	18.7	0	18.7	85.(7.)
^{77}Se	7.6	8.25	(+/-)0.6(1.6)	8.6	0.05	8.65	42.(4.)
^{78}Se	23.5	8.24	0	8.5	0	8.5	0.43
^{80}Se	49.6	7.48	0	7.03	0	7.03	0.61
^{82}Se	9.4	6.34	0	5.05	0	5.05	0.044
Br	—	6.795	—	5.8	0.1	5.9	6.9
^{79}Br	50.69	6.80	-1.1	5.81	0.15	5.96	11

(note: 1 fm = 10^{-15} m, 1 barn = 10^{-24} cm^2)

TABLE G.1: (continued)

Isotope	conc (%)	b_{coh} (fm)	b_{inc} (fm)	σ_{coh}	σ_{inc}	σ_{tot}	σ_{abs}
^{81}Br	49.31	6.79	0.6	5.79	0.05	5.84	2.7
Kr	—	7.81	—	7.67	0.01	7.68	25.(1.)
^{78}Kr	0.35	—	0	—	0	—	6.4
^{80}Kr	2.25	—	0	—	0	—	11.8
^{82}Kr	11.6	—	0	—	0	—	29.(20.)
^{83}Kr	11.5	—	—	—	—	—	185.(30.)
^{84}Kr	57	—	0	—	0	6.6	0.113
^{86}Kr	17.3	8.1	0	8.2	0	8.2	0.003
Rb	—	7.09	—	6.32	0.5	6.8	0.38
^{85}Rb	72.17	7.03	—	6.2	0.5	6.7	0.48
^{87}Rb	27.83	7.23	—	6.6	0.5	7.1	0.12
Sr	—	7.02	—	6.19	0.06	6.25	1.28
^{84}Sr	0.56	7.(1.)	0	6.(2.)	0	6.(2.)	0.87
^{86}Sr	9.86	5.67	0	4.04	0	4.04	1.04
^{87}Sr	7	7.40	—	6.88	0.5	7.4	16.(3.)
^{88}Sr	82.58	7.15	0	6.42	0	6.42	0.058
Y	100	7.75	1.1	7.55	0.15	7.7	1.28
Zr	—	7.16	—	6.44	0.02	6.46	0.185
^{90}Zr	51.45	6.4	0	5.1	0	5.1	0.011
^{91}Zr	11.32	8.7	-1.08	9.5	0.15	9.7	1.17
^{92}Zr	17.19	7.4	0	6.9	0	6.9	0.22
^{94}Zr	17.28	8.2	0	8.4	0	8.4	0.0499
^{96}Zr	2.76	5.5	0	3.8	0	3.8	0.0229
Nb	100	7.054	-0.139	6.253	0.0024	6.255	1.15
Mo	—	6.715	—	5.67	0.04	5.71	2.48
^{92}Mo	14.84	6.91	0	6	0	6	0.019
^{94}Mo	9.25	6.80	0	5.81	0	5.81	0.015
^{95}Mo	15.92	6.91	—	6	0.5	6.5	13.1
^{96}Mo	16.68	6.20	0	4.83	0	4.83	0.5
^{97}Mo	9.55	7.24	—	6.59	0.5	7.1	2.5
^{98}Mo	24.13	6.58	0	5.44	0	5.44	0.127
^{100}Mo	9.63	6.73	0	5.69	0	5.69	0.4
Tc	(2.3E5 a)	6.8	—	5.8	0.5	6.3	20.(1.)
Ru	—	7.03	—	6.21	0.4	6.6	2.56
^{96}Ru	5.5	—	0	—	0	—	0.28
^{98}Ru	1.9	—	0	—	0	—	¡8.
^{99}Ru	12.7	—	—	—	—	—	6.9(1.0)
^{100}Ru	12.6	—	0	—	0	—	4.8
^{101}Ru	17	—	—	—	—	—	3.3
^{102}Ru	31.6	—	0	—	0	144.8	1.17
^{104}Ru	18.7	—	0	—	0	4.483	0.31
Rh	100	5.88	—	4.34	0.3	4.6	144.8
Pd	—	5.91	—	4.39	0.093	4.48	6.9
^{102}Pd	1.02	7.7(7)	0	7.5(1.4)	0	7.5(1.4)	3.4
^{104}Pd	11.14	7.7(7)	0	7.5(1.4)	0	7.5(1.4)	0.6
^{105}Pd	22.33	5.5	-2.6(1.6)	3.8	0.8	4.6(1.1)	20.(3.)
^{106}Pd	27.33	6.4	0	5.1	0	5.1	0.304
^{108}Pd	26.46	4.1	0	2.1	0	2.1	8.55
^{110}Pd	11.72	7.7(7)	0	7.5(1.4)	0	7.5(1.4)	0.226
Ag	—	5.922	—	4.407	0.58	4.99	63.3
^{107}Ag	51.83	7.555	1.00	7.17	0.13	7.3	37.6(1.2)
^{109}Ag	48.17	4.165	-1.60	2.18	0.32	2.5	91.0(1.0)
Cd	—	4.87-0.70i	—	3.04	3.46	6.5	2520.(50.)
^{106}Cd	1.25	5.(2.)	0	3.1	0	3.1(2.5)	1
^{108}Cd	0.89	5.4	0	3.7	0	3.7	1.1
^{110}Cd	12.51	5.9	0	4.4	0	4.4	11
^{111}Cd	12.81	6.5	—	5.3	0.3	5.6	24
^{112}Cd	24.13	6.4	0	5.1	0	5.1	2.2
^{113}Cd	12.22	-8.0-5.73i	—	12.1	0.3	12.4	20600.(400.)
^{114}Cd	28.72	7.5	0	7.1	0	7.1	0.34
^{116}Cd	7.47	6.3	0	5	0	5	0.075
In	—	4.065-0.0539i	—	2.08	0.54	2.62	193.8(1.5)
^{113}In	4.3	5.39	(+/-)0.017	3.65	0.000037	3.65	12.0(1.1)
^{115}In	95.7	4.01-0.0562i	-2.1	2.02	0.55	2.57	202.(2.)
Sn	—	6.225	—	4.871	0.022	4.892	0.626
^{112}Sn	1	6.(1.)	0	4.5(1.5)	0	4.5(1.5)	1
^{114}Sn	0.7	6.2	0	4.8	0	4.8	0.114

(note: 1 fm $= 10^{-15}$ m, 1 barn $= 10^{-24}$ cm^2)

TABLE G.1: (continued)

Isotope	conc (%)	b_{coh} (fm)	b_{inc} (fm)	σ_{coh}	σ_{inc}	σ_{tot}	σ_{abs}
^{115}Sn	0.4	6.(1.)	—	4.5(1.5)	0.3	4.8(1.5)	30.(7.)
^{116}Sn	14.7	5.93	0	4.42	0	4.42	0.14
^{117}Sn	7.7	6.48	—	5.28	0.3	5.6	2.3
^{118}Sn	24.3	6.07	0	4.63	0	4.63	0.22
^{119}Sn	8.6	6.12	—	4.71	0.3	5	2.2
^{120}Sn	32.4	6.49	0	5.29	0	5.29	0.14
^{122}Sn	4.6	5.74	0	4.14	0	4.14	0.18
^{124}Sn	5.6	5.97	0	4.48	0	4.48	0.133
Sb	—	5.57	—	3.9	0.007	3.9	4.91
^{121}Sb	57.0	5.71	-0.05	4.1	0.0003	4.1	5.75
^{123}Sb	42.7	5.38	-0.10	3.64	0.001	3.64	3.8
Te	—	5.80	—	4.23	0.09	4.32	4.7
^{120}Te	0.096	5.3	0	3.5	0	3.5	2.3
^{122}Te	2.6	3.8	0	1.8	0	1.8	3.4
^{123}Te	0.908	-0.05-0.116i	-2.04	0.002	0.52	0.52	418.(30.)
^{124}Te	4.816	7.96	0	8	0	8	6.8(1.3)
^{125}Te	7.14	5.02	-0.26	3.17	0.008	3.18	1.55
^{126}Te	18.95	5.56	0	3.88	0	3.88	1.04
^{128}Te	31.69	5.89	0	4.36	0	4.36	0.215
^{130}Te	33.8	6.02	0	4.55	0	4.55	0.29
I	100	5.28	1.58	3.5	0.31	3.81	6.15
Xe	—	4.92	3.04	2.96	0	—	23.9(1.2)
^{124}Xe	0.1	—	0	—	0	—	165.(20.)
^{126}Xe	0.09	—	0	—	0	—	3.5
^{128}Xe	1.91	—	0	—	0	—	¡8
^{129}Xe	26.4	—	—	—	—	—	21.(5.)
^{130}Xe	4.1	—	0	—	0	—	¡26.
^{131}Xe	21.2	—	—	—	—	—	85.(10.)
^{132}Xe	26.9	—	0	—	0	—	0.45
^{134}Xe	10.4	—	0	—	0	—	0.265
^{136}Xe	8.9	—	0	—	0	—	0.26
Cs	100	5.42	1.29	3.69	0.21	3.9	29.0(1.5)
Ba	—	5.07	—	3.23	0.15	3.38	1.1
^{130}Ba	0.11	-3.6	0	1.6	0	1.6	30.(5.)
^{132}Ba	0.1	7.8	0	7.6	0	7.6	7
^{134}Ba	2.42	5.7	0	4.08	0	4.08	2.0(1.6)
^{135}Ba	6.59	4.67	—	2.74	0.5	3.2	5.8
^{136}Ba	7.85	4.91	0	3.03	0	3.03	0.68
^{137}Ba	11.23	6.83	—	5.86	0.5	6.4	3.6
^{138}Ba	71.7	4.84	0	2.94	0	2.94	0.27
La	—	8.24	—	8.53	1.13	9.66	8.97
^{138}La	0.09	8.(2.)	—	8.(4.)	0.5	8.5(4.0)	57.(6.)
^{139}La	99.91	8.24	3.0	8.53	1.13	9.66	8.93
Ce	—	4.84	—	2.94	0.001	2.94	0.63
^{136}Ce	0.19	5.80	0	4.23	0	4.23	7.3(1.5)
^{138}Ce	0.25	6.70	0	5.64	0	5.64	1.1
^{140}Ce	88.48	4.84	0	2.94	0	2.94	0.57
^{142}Ce	11.08	4.75	0	2.84	0	2.84	0.95
Pr	100	4.58	-0.35	2.64	0.015	2.66	11.5
Nd	—	7.69	—	7.43	9.2	16.6	50.5(1.2)
^{142}Nd	27.16	7.7	0	7.5	0	7.5	18.7
^{143}Nd	12.18	14.(2.)	(+/-)21.(1.)	25.(7.)	55.(7.)	80.(2.)	337.(10.)
^{144}Nd	23.8	2.8	0	1	0	1	3.6
^{145}Nd	8.29	14.(2.)	—	25.(7.)	5.(5.)	30.(9.)	42.(2.)
^{146}Nd	17.19	8.7	0	9.5	0	9.5	1.4
^{148}Nd	5.75	5.7	0	4.1	0	4.1	2.5
^{150}Nd	5.63	5.3	0	3.5	0	3.5	1.2
Pm	(2.62 a)	12.6	(+/-)3.2(2.5)	20.0(1.3)	1.3(2.0)	21.3(1.5)	168.4(3.5)
Sm	—	0.80-1.65i	—	0.422	39.(3.)	39.(3.)	5922.(56.)
^{144}Sm	3.1	-3.(4.)	0	1.(3.)	0	1.(3.)	0.7
^{147}Sm	15.1	14.(3.)	(+/-)11.(7.)	25.(11.)	143(19.)	39.(16.)	57.(3.)
^{148}Sm	11.3	-3.(4.)	0	1.(3.)	0	1.(3.)	2.4
^{149}Sm	13.9	-19.2-11.7i	(+/-)31.4-10.3i	63.5	137.(5.)	200.(5.)	42080.(400.)
^{150}Sm	7.4	14.(3.)	0	25.(11.)	0	25.(11.)	104.(4.)
^{152}Sm	26.6	-5.0	0	3.1	0	3.1	206.(6.)
^{154}Sm	22.6	9.3	0	11.(2.)	0	11.(2.)	8.4
Eu	—	7.22-1.26i	—	6.57	2.5	9.2	4530.(40.)

(note: 1 fm = 10^{-15} m, 1 barn = 10^{-24} cm^2)

TABLE G.1: (continued)

Isotope	conc (%)	b_{coh} (fm)	b_{inc} (fm)	σ_{coh}	σ_{inc}	σ_{tot}	σ_{abs}
^{151}Eu	47.8	6.13-2.53i	(+/-)4.5-2.14i	5.5	3.1	8.6	9100.(100.)
^{153}Eu	52.2	8.22	(+/-)3.2	8.5	1.3	9.8	312.(7.)
Gd	—	6.5-13.82i	—	29.3	151.(2.)	180.(2.)	49700.(125.)
^{152}Gd	0.2	10.(3.)	0	13.(8.)	0	13.(8.)	735.(20.)
^{154}Gd	2.1	10.(3.)	0	13.(8.)	0	13.(8.)	85.(12.)
^{155}Gd	14.8	6.0-17.0i	(+/-)5.(5.)-13.16i	40.8	25.(6.)	66.(6.)	61100.(400.)
^{156}Gd	20.6	6.3	0	5	0	5	1.5(1.2)
^{157}Gd	15.7	-1.14-71.9i	(+/-)5.(5.)-55.8i	650.(4.)	394.(7.)	1044.(8.)	259000.(700.)
^{158}Gd	24.8	9.(2.)	0	10.(5.)	0	10.(5.)	2.2
^{160}Gd	21.8	9.15	0	10.52	0	10.52	0.77
Tb	100	7.38	-0.17	6.84	0.004	6.84	23.4
Dy		16.9-0.276i	—	35.9	54.4(1.2)	90.3	994.(13.)
^{156}Dy	0.06	6.1	0	4.7	0	4.7	33.(3.)
^{158}Dy	0.1	6.(4.)	0	5.(6.)	0	5.(6.)	43.(6.)
^{160}Dy	2.34	6.7	0	5.6	0	5.6	56.(5.)
^{161}Dy	19	10.3	(+/-)4.9	13.3	3.(1.)	16.(1.)	600.(25.)
^{162}Dy	25.5	-1.4	0	0.25	0	0.25	194.(10.)
^{163}Dy	24.9	5.0	1.3	3.1	0.21	3.3	124.(7.)
^{164}Dy	28.1	49.4-0.79i	0	307.(3.)	0	307.(3.)	2840.(40.)
Ho	100	8.01	-1.70	8.06	0.36	8.42	64.7(1.2)
Er	—	7.79	—	7.63	1.1	8.7	159.(4.)
^{162}Er	0.14	8.8	0	9.7	0	9.7	19.(2.)
^{164}Er	1.56	8.2	0	8.4	0	8.4	13.(2.)
^{166}Er	33.4	10.6	0	14.1	0	14.1	19.6(1.5)
^{167}Er	22.9	3.0	1.0	1.1	0.13	1.2	659.(16.)
^{168}Er	27.1	7.4	0	6.9	0	6.9	2.74
^{170}Er	14.9	9.6	0	11.6	0	11.6(1.2)	5.8
Tm	100	7.07	0.9	6.28	0.1	6.38	100.(2.)
Yb	—	12.43		19.42	4	23.4	34.8
^{168}Yb	0.14	-4.07-0.62i	0	2.13	0	2.13	2230 (40.)
^{170}Yb	3.06	6.77	0	5.8	0	5.8	11.4(1.0)
^{171}Yb	14.3	9.66	-5.59	11.7	3.9	15.6	48.6(2.5)
^{172}Yb	21.9	9.43	0	11.2	0	11.2	0.8
^{173}Yb	16.1	9.56	-5.3	11.5	3.5	15	17.1(1.3)
^{174}Yb	31.8	19.3	0	46.8	0	46.8	69.4(5.0)
^{176}Yb	12.7	8.72	0	9.6	0	9.6	2.85
Lu	—	7.21	—	6.53	0.7	7.2	74.(2.)
^{175}Lu	97.39	7.24	(+/-)2.2	6.59	0.6	7.2	21.(3.)
^{176}Lu	2.61	6.1-0.57i	(+/-)3.0+0.61i	4.7	1.2	5.9	2065.(35.)
Hf	—	7.7	—	7.6	2.6	10.2	104.1
^{174}Hf	0.2	10.9(1.1)	0	15.(3.)	0	15.(3.)	561.(35.)
^{176}Hf	5.2	6.61	0	5.5	0	5.5	23.5(3.1)
^{177}Hf	18.6	0.8(1.0)	(+/-)0.9(1.3)	0.1	0.1	0.2	373.(10.)
^{178}Hf	27.1	5.9	0	4.4	0	4.4	84.(4.)
^{179}Hf	13.7	7.46	(+/-)1.06	7	0.14	7.1	41.(3.)
^{180}Hf	35.2	13.2	0	21.9	0	21.9(1.0)	13.04
Ta	—	6.91	—	6	0.01	6.01	20.6
^{180}Ta	0.012	7.(2.)	—	6.2	0.5	7.(4.)	563.(60.)
^{181}Ta	99.988	6.91	-0.29	6	0.011	6.01	20.5
W	—	4.86	—	2.97	1.63	4.6	18.3
^{180}W	0.1	5.(3.)	0	3.(4.)	0	3.(4.)	30.(20.)
^{182}W	26.3	6.97	0	6.1	0	6.1	20.7
^{183}W	14.3	6.53	—	5.36	0.3	5.7	10.1
^{184}W	30.7	7.48	0	7.03	0	7.03	1.7
^{186}W	28.6	-0.72	0	0.065	0	0.065	37.9
Re	—	9.2	—	10.6	0.9	11.5	89.7(1.)
^{185}Re	37.4	9.0	(+/-)2.0	10.2	0.5	10.7	112.(2.)
^{187}Re	62.6	9.3	(+/-)2.8	10.9	1	11.9	76.4(1.)
Os	—	10.7	—	14.4	0.3	14.7	16
^{184}Os	0.02	10.(2.)	0	13.(5.)	0	13.(5.)	3000.(150.)
^{186}Os	1.58	11.6(1.7)	0	17.(5.)	0	17.(5.)	80.(13.)
^{187}Os	1.6	10.(2.)	—	13.(5.)	0.3	13.(5.)	320.(10.)
^{188}Os	13.3	7.6	0	7.3	0	7.3	4.7
^{189}Os	16.1	10.7	—	14.4	0.5	14.9	25.(4.)
^{190}Os	26.4	11.0	0	15.2	0	15.2	13.1
^{192}Os	41	11.5	0	16.6	0	16.6(1.2)	2
Ir	—	10.6	—	14.1	0.(3.)	14.(3.)	425.(2.)

(note: 1 fm = 10^{-15} m, 1 barn = 10^{-24} cm^2)

TABLE G.1: (continued)

Isotope	conc (%)	b_{coh} (fm)	b_{inc} (fm)	σ_{coh}	σ_{inc}	σ_{tot}	σ_{abs}
^{191}Ir	37.3	—	—	—	—	—	954.(10.)
^{193}Ir	62.7	—	—	—	—	—	111.(5.)
Pt	—	9.60	—	11.58	0.13	11.71	10.3
^{190}Pt	0.01	9.0	0	10.(2.)	0	10.(2.)	152.(4.)
^{192}Pt	0.79	9.9	0	12.3(1.2)	0	12.3(1.2)	10.0(2.5)
^{194}Pt	32.9	10.55	0	14	0	14	1.44
^{195}Pt	33.8	8.83	-1.00	9.8	0.13	9.9	27.5(1.2)
^{196}Pt	25.3	9.89	0	12.3	0	12.3	0.72
^{198}Pt	7.2	7.8	0	7.6	0	7.6	3.66
Au	100	7.63	-1.84	7.32	0.43	7.75	98.65
Hg	—	12.692	—	20.24	6.6	26.8	372.3(4.0)
^{196}Hg	0.2	30.3(1.0)	0	115.(8.)	0	115.(8.)	3080.(180.)
^{198}Hg	10.1	—	0	—	0	—	2
^{199}Hg	17	16.9	(+/-)15.5	36.(2.)	30.(3.)	66.(2.)	2150.(48.)
^{200}Hg	23.1	—	0	—	0	—	¡60.
^{201}Hg	13.2	—	—	—	-.-	—	7.8(2.0)
^{202}Hg	29.6	—	0	—	0	9.828	4.89
^{204}Hg	6.8	—	0	—	0	—	0.43
Tl	—	8.776	—	9.678	0.21	9.89	3.43
^{203}Tl	29.524	6.99	1.06	6.14	0.14	6.28	11.4
^{205}Tl	70.476	9.52	-0.242	11.39	0.007	11.4	0.104
Pb	—	9.405	—	11.115	0.003	11.118	0.171
^{204}Pb	1.4	9.90	0	12.3	0	12.3	0.65
^{206}Pb	24.1	9.22	0	10.68	0	10.68	0.03
^{207}Pb	22.1	9.28	0.14	10.82	0.002	10.82	0.699
^{208}Pb	52.4	9.50	0	11.34	0	11.34	0.00048
Bi	100	8.532	—	9.148	0.0084	9.156	0.0338
Po	—	—	0.259	0	—	—	—
At	—	—	—	0	—	—	—
Rn	—	—	—	0	—	12.6	—
Fr	—	—	—	0	—	—	—
Ra	(1.60E3 a)	10.0(1.0)	0	13.(3.)	0	13.(3.)	12.8(1.5)
Ac	—	—	—	0	—	—	—
Th	100	10.31	0	13.36	0	13.36	7.37
Pa	(3.28E4 a)	9.1	—	10.4	0.1(3.3)	10.5(3.2)	200.6(2.3)
U	—	8.417	—	8.903	0.005	8.908	7.57
^{233}U	(1.59E5 a)	10.1	(+/-)1.(3.)	12.8	0.1	12.9	574.7(1.0)
^{234}U	0.005	12.4	0	19.3	0	19.3	100.1(1.3)
^{235}U	0.72	10.47	(+/-)1.3	13.78	0.2	14	680.9(1.1)
^{238}U	99.275	8.402	0	8.871	0	8.871	2.68
Np	(2.14E6 a)	10.55	—	14	0.5	14.5	175.9(2.9)
Pu	—	—	—	—	—	—	—
^{238}Pu	(87.74 a)	14.1	0	25.0(1.8)	0	25.0(1.8)	558.(7.)
^{239}Pu	(2.41E4 a)	7.7	(+/-)1.3(1.9)	7.5	0.2	7.7	1017.3(2.1)
^{240}Pu	(6.56E3 a)	3.5	0	1.54	0	1.54	289.6(1.4)
^{242}Pu	(3.76E5 a)	8.1	0	8.2	0	8.2	18.5
Am	(7.37E3 a)	8.3	(+/-)2.(7.)	8.7	0.3	9.0(2.6)	75.3(1.8)
Cm	—	—	—	0	—	—	—
^{244}Cm	(18.10 a)	9.5	0	11.3	0	11.3	16.2(1.2)
^{246}Cm	(4.7E3 a)	9.3	0	10.9	0	10.9	1.36
^{248}Cm	(3.5E5 a)	7.7	0	7.5	0	7.5	3

(note: 1 fm = 10^{-15} m, 1 barn = 10^{-24} cm^2)

Appendix H

Activation table of elements

It is useful to know if a high pressure equipment activates (i.e. becomes radioactive) after it has been exposed to a neutron beam. The choice of the construction materials has to consider this possibility and to avoid activation as much as possible. A useful aid in estimating the risk of activation is the table given below. It is essentially a subset of data collected by S. Mason which can be found on most sites of neutron facilities. This table has three essential entrees:

1. The *storage time* which is the time required for a given element in condensed form exposed to a "standard" neutron beam to decay to 2 nCi/g or less. This is the limit for shipping a material as "nonradioactive".

2. The *prompt activation* which gives the anticipated activation for a pure solid element two minutes after the neutron exposure has stopped.

3. The *contact dose* for a 1 g sample of the pure element due to the prompt activation. Elements with a '-' for the entries in all three columns do not show any activation. Bismuth is a special case; it is stable before exposure to the beam, but the activation product is an α-emitter.

Using these numbers it is in principle possible to calculate the activation of any compound if the flux of the beam compared to the "standard" beam is known. The reader is referred to the relevant web sites for examples.

TABLE H.1: Neutron activation table for selected elements, after S. Mason. m = minutes, h= hours, d= days, y= years. Source: ILL.

Symbol	Name	Mass	Storage time	Prompt activation[a]	Contact dose[b]
Al	aluminium	26.982	21m	1900	2.0
Sb	antimony	121.75	520d	800	0.7
Ar	argon	39.948	19h	3500	3.0
As	arsenic	74.922	18d	8.4×10^4	7.3
Ba	barium	137.34	<150h	<80	<0.1
Be	beryllium	9.012	-	-	-
Bi	bismuth	209.98	**	**	**
B	boron	10.81	-	-	-
Br	bromine	79.909	18d	1.4×10^4	12
Cd	cadmium	112.4	190d	370	0.3
Ca	calcium	40.08	-	-	-
C	carbon	12.01	-	-	-

[a] in nCi/g = 37 Bq/g; [b] ini mr/hr/g = 10μ Sv/hr/g at 2.5 cm distance.

TABLE H.1: (continued)

Symbol	Name	Mass	Storage time	Prompt activation[a]	Contact dose[b]
Ce	cerium	14.12	<86h	<40	<0.1
Cs	cesium	132.905	54h	4.6×10^5	400
Cl	chlorine	35.453	<2.8h	<80	<0.1
Cr	chromium	51.996	<6ld	<40	<0.1
Co	cobalt	58.933	24y	5.2×10^4	45
Cu	copper	73.54	7.4d	1.0×10^4	8.5
Dy	dysprosium	162.5	52h	5.0×10^5	430
D	deuterium	2.015	-	-	-
Er	erbium	167.26	78d	600	0.5
Eu	europium	151.96	50y	2200	1.9
F	fluorine	18.998	-	-	-
Gd	gadolinium	157.25	11d	7400	6.d
Ga	gallium	69.72	8d	3.2×10^4	27
Ge	germanium	72.59	<6d	1100	1.0
Au	gold	196.967	29d	3000	2.5
Hf	hafnium	178.49	1.6y	620	0.5
He	helium	4.003	-	-	-
Ho	holmium	164.93	20d	2.8×10^4	24
H	hydrogen	1.008	-	-	-
In	indium	114.82	12d	1.1×10^4	9.5
I	iodine	126.904	7h	1.2×10^5	100
Ir	iridium	192.2	4.2y	5.0×10^4	43
Fe	iron	55.847	-	-	-
Kr	krypton	83.80	42h	3200	2.8
La	lanthanum	138.91	22d	1.9×10^4	16
Pb	lead	207.19	-	-	-
Li	lithium	6.939	-	-	-
Lu	lutetium	174.97	1.8y	1.4×10^4	12
Mg	magnesium	24.312	-	-	-
Mn	manganese	54.938	38h	1.1×10^5	95
Hg	mercury	200.59	24d	700	0.6
Mo	molybdenum	95.94	30d	430	0.4
Nd	neodymium	144.24	15h	1200	1.0
Ne	neon	20.183	-	-	-
Ni	nickel	58.71	<5.5h	<30	<0.1
Nb	niobium	92.906	80m	2.0×10^4	17
N	nitrogen	14.007	-	-	-
Os	osmium	190.2	41d	2300	2.0
O	oxygen	15.999	-	-	-
Pd	palladium	160.4	9d	7.1×10^4	60
P	phosphorus	30.974	-	-	-
Pt	platinum	195.09	20d	230	0.2
K	potassium	39.102	<38h	<300	<0.3
Pr	praseodymium	140.907	11d	2.0×10^4	17

[a] in nCi/g = 37 Bq/g; [b] ini mr/hr/g = 10μ Sv/hr/g at 2.5 cm distance.

TABLE H.1: (continued)

Symbol	Name	Mass	Storage time	Prompt activation[a]	Contact dose[b]
Re	rhenium	186.2	53d	4.9×10^4	42
Rh	rhodium	102.905	2h	2.6×10^4	22
Rb	rubidium	85.47	56d	1800	1.6
Ru	ruthenium	101.07	106d	230	0.2
Sm	samarium	150.35	35d	6200	5.4
Sc	scandium	44.956	<1.8y	<90	<0.1
Se	selenium	78.96	10h	4900	4.2
Si	silicon	28.086	-	-	-
Ag	silver	107.87	7.4y	1.6×10^4	14
Na	sodium	22.991	5.5d	5700	5.0
Sr	strontium	87.62	<25h	< 100	<100
Pd	palladium	160.4	9d	7.1×10^4	60
S	sulphur	32.064	-	-	-
Ta	tantalum	180.948	3y	1600	1.4
Te	tellurium	127.6	96h	2600	2.2
Tb	terbium	158.924	2.1y	3300	2.8
Tl	thallium	204.37	41m	460	0.4
Tm	thulium	168.934	3.7y	7700	6.7
Sn	tin	118.69	<50d	<40	<0.1
Ti	titanium	47.90	-	-	-
W	tungsten	183.85	15d	3.7×10^4	32
V	vanadium	50.942	48m	4.7×10^5	41
Xe	xenon	131.30	7d	3200	2.8
Yb	ytterbium	173.04	275d	780	0.7
Y	yttrium	88.905	24d	1000	0.9
Zn	zinc	65.37	5d	1600	1.4
Zr	zirconium	91.22	79h	<40	<0.1

[a] in nCi/g = 37 Bq/g; [b] ini mr/hr/g = 10μ Sv/hr/g at 2.5 cm distance.

Appendix I

Thermal expansion of Pb and NaCl at low temperatures

TABLE I.1

Thermal volume expansion of Pb and NaCl relative to T=300 K (V_0).
Typical reference values at 300 K are V_0=121.29 $\mathring{A}^3$ (a=4.950 $\mathring{A}$) for Pb and
V_0=179.41 $\mathring{A}^3$ (a=5.640 $\mathring{A}$) for NaCl. Data were obtained by neutron
diffraction on a Pb/NaCl powder mixture (Strässle et al., unpublished.)

T(K)	V/V_0(Pb)	V/V_0(NaCl)	T(K)	V/V_0(Pb)	V/V_0(NaCl)
0	0.9783	0.9761	160	0.9884	0.9846
10	0.9784	0.9762	170	0.9892	0.9856
20	0.9786	0.9762	180	0.9900	0.9866
30	0.9789	0.9763	190	0.9908	0.9877
40	0.9794	0.9764	200	0.9916	0.9887
50	0.9800	0.9766	210	0.9924	0.9898
60	0.9807	0.9769	220	0.9933	0.9909
70	0.9814	0.9773	230	0.9941	0.9921
80	0.9822	0.9778	240	0.9949	0.9932
90	0.9829	0.9785	250	0.9958	0.9943
100	0.9837	0.9792	260	0.9966	0.9954
110	0.9845	0.9800	270	0.9975	0.9966
120	0.9852	0.9809	280	0.9983	0.9977
130	0.9860	0.9818	290	0.9992	0.9989
140	0.9867	0.9827	300	1.0000	1.0000
150	0.9876	0.9836			

Bibliography

[1] J.J. Kruzic, D.K. Kim, K.J. Koester, and R.O. Ritchie. Indentation techniques for evaluating the fracture toughness of biomaterials and hard tissues. *J. Mech. Beh. Biomed. Mat.*, 2:384–395, 2009.

[2] J.-P. Poirier. *Introduction to the Physics of the Earth's Interior, 2nd Edition*. Cambridge University Press, 2000.

[3] I.L. Spain and J. Paauwe (edts.). *High Pressure Technology, Vol. I: Equipment Design, Materials, and Properties*. Marcel Dekker Inc., New York/Basel, 1977.

[4] M. Fremets. *High Pressure Experimental Methods*. Oxford University Press, Oxford, 1996.

[5] W.F. Sherman and A.A. Stadtmuller. *Experimental Techniques in High-Pressure Research*. John Wiley & Sons, Ltd. New York, NY, USA, 1987.

[6] J. Paureau and Ch. Vettier. New high pressure cell for neutron scattering at very low temperature. *Rev. Sci. Instrum.*, 46:1484–1488, 1975.

[7] O. Blaschko and G. Ernst. Autofrettaged high pressure chamber for use in inelastic neutron scattering. *Rev. Sci. Instrum.*, 45:526–528, 1974.

[8] C. Cooley and M. Aronson. Origns of paramagnetism in beryllium-copper. *J. Alloys Compds.*, 228:195–200, 1995.

[9] S.S. Sidhu, L. Heaton, D.D. Zauberis, and F.P. Campos. Neutron diffraction study of titanium-zirconium system. *J. Appl. Phys.*, 27:1040–1042, 1956.

[10] J. Zhang et al. Experimental constraints on the phase diagram of elemental zirconium. *J. Phys. Chem. Solids*, 66:1213–1219, 2005.

[11] H. Xia, J. Duclos, A.L. Ruoff, and Y.K. Vohra. New high-pressure phase transition in zirconium metal. *Phys. Rev. Lett.*, 64:204–207, 1990.

[12] H. Xia, H. Parthasarathy, G. Luo, Y.K. Vohra, and A.L. Ruoff. Crystal structures of group IVa metals at ultrahigh pressures. *Phys. Rev. B*, 42:204–207, 1990.

[13] I.O. Bashkin et al. Crystal structure and superconductivity of TiZr up to 57 GPa. *Phys. Rev. B*, 68:054401–6, 2003.

[14] V.P. Dimitriev et al. Collapsed hexagonal ω phase in a compressed TiZr alloy: Angle-dispersive synchrotron-radiation x-ray diffraction study. *Phys. Rev. B.*, 73:094114–6, 2006.

[15] C. Wegst and M. Wegst. *Stahlschlüssel - Key to Steel - La Clé des Aciers, 22nd edition*. Verlag Stahlschlüssel West GmbH, Germany, 2010.

[16] K. Vetter, E. Gondolf, and A. von der Steinen. *Hochfeste Stähle. In: Werkstoffkunde Stahl, Band 2: Anwendungen, edtrs. W. Jäniche et al.,*. Springer Verlag, 1985.

[17] I.R. Walker. Considerations on the selection of alloys for use in pressure cells at low temperatures. *Cryogenics*, 45:87–108, 2005.

[18] Y. Uwatoko et al. Material properties of Ni-Cr-Al alloy and design of a 4 GPa class non-magnetic high-pressure cell. *J. Phys.: Condens. Matter*, 14:11291–11296, 2002.

[19] G.T. Murray. In *Hydrogen Embrittlement: Prevention and Control, L. Raymond edt., ASTM special techn. publ. no. 962*, pages 304–317. ASTM, Philadelphia.

[20] J. Bowker and G.R. Piercy. The effect of a tin barrier layer on the permeability of hydrogen through mild steel and ferritic stainless steel. *Metall. Transactions A*, 15A:2093–2095, 1984.

[21] G.M. Pressouyre. In *Proc. of 1st Int. Conf. on Current Solutions of Hydrogen Problems in Steel, C.G. Interrante and G.M. Pressouyre edts.*, pages 18–34. American Soc. for Metals, Metals Park, OH.

[22] M.D. Horton, B.J. Pope, L.B. Horton, and R.P. Radtke. Some properties of sintered diamond. In *High Pressure Science and Technology*, volume 1, pages 923–930. Plenum Press, New York, 1979.

[23] W. Utsumi, T. Yagi, K. Leinenweber, O. Shimomura, and T. Taniguchi. *High pressure and high temperature generation using sintered diamond anvils*. Terra Scientific Publishing Company, Amercian Geophysical Union, 1992.

[24] T. Irifune, A. Kurio, S. Sakamoto, T. Inoue, and H. Sumiya. Ultrahard polycrystalline diamond from graphite. *Nature*, 421:599–600, 2003.

[25] Y. Nakamoto et al. Note: High-pressure generation using nano-polycrystalline diamonds as anvil materials. *Rev. Sci. Instrum.*, 82:066104–2, 2011.

[26] J. Peters et al. High hydrostatic pressure equipment for neutron scattering studies of samples in solution. *High Pressure Res.*, 31:97–102, 2012.

[27] M.S. Appavou et al. Using pressure in combination with x-ray and neutron scattering techniques for studying the structure, stability and phase behaviour of soft condensed matter and biomolecular systems. *J. Phys.: Condens. Matter*, 17:S3093–S3099, 2005.

[28] S. Susman and R. Kleb. A new high pressure apparatus for neutron diffraction in liquids and glasses at 0.7 GPa. *High Pressure Res.*, 11:263–269, 1993.

[29] D.B. McWhan, D. Bloch, and G. Parisot. Apparatus for neutron diffraction at high pressure. *Rev. Sci. Instrum.*, 45:643–646, 1974.

[30] R.M. Brugger, R.B. Bennion, and T.G. Worlton. The crystal structure of Bismuth-II at 26 kbar. *Physics Letters*, 24:714–717, 1967.

[31] T.G. Worlton and D.L. Decker. Neutron diffraction study of the magnetic structure of hematite to 41 kbar. *Phys. Rev.*, 171:596–599, 1968.

[32] D. Bloch, J. Paureau, J. Voiron, and G. Parisot. Neutron scattering at high pressure. *Rev. Sci. Instrum.*, 47:296–298, 1976.

[33] D.B. McWhan and C. Vettier. Neutron scattering studies at high pressure on rare earth intermetallic compounds. *J. Physique - Colloques*, C5:107–111, 1979.

[34] D.B. McWhan, C. Vettier, R. Youngblood, and G. Shirane. Neutron scattering studies of pressure-induced antiferromagnetism in PrSb. *Phys. Rev. B*, 20:4612–4623, 1979.

[35] D. Bloch and J. Paureau. In *High Pressure Chemistry, Proc. NATO Adv. Study Inst. held in Corfu, Greece, sept. 1977; edited by H. Kelm.*, pages 111–126. D. Reidel Publishing Company.

[36] I.F. Bailey. A review of sample environments in neutron scattering. *Z. Kristallogr.*, 218:84–95, 2003.

[37] R.G. Munro. Evaluated material properties for a sintered α-alumina. *J. Am. Ceram. Soc.*, 80:1919–1928, 1997.

[38] U. Walter. A pressure cell for inelastic neutron scattering up to 35 kbar. *Revue Phys. Appl.*, 19:833–836, 1984.

[39] W.F. Kuhs, J.L. Finney, C. Vettier, and D.V. Bliss. Structure and hydrogen ordering in ices VI, VII, and VIII by neutron powder diffraction. *J. Chem. Phys*, 81:3612–3623, 1984.

[40] J.B. Forsyth and S. Hull. The effect of hydrostatic pressure on the ambient temperature structure of CuO. *J. Phys.: Condens. Matter*, 3:5257–5261, 1991.

[41] S. Hull and D.A. Keen. Structural modifications within copper(I) Iodine at pressures up to 27 kbar. *Europhys. Lett.*, 23:129–134, 1993.

[42] J.E. Ibballs, R.J. Nelmes, and G.J. McIntyre. The crystral structure of tetragonal KH_2PO_4 and KD_2PO_4 as a function of temperature and presssure. *J. Phys. C: Solid State Phys.*, 15:37–58, 1982.

[43] A. Onodera et al, High-pressure cell for neutron scattering. *Jpn. J. Appl. Phys.*, 26:152–156, 1987.

[44] K. Knorr, K. Fütterer, B. Annighöfer, and W. Depmeier. A heatable large volume high pressure cell for neutron powder diffraction: The Kiel-Berlin Cell I. *Rev. Sci. Instrum.*, 68:3817–3822, 1997.

[45] S. Hull, D.A. Keen, R. Done, T. Pike, and N.J.G. Gardner. A high temperature, high pressure cell for time-of-flight neutron scattering. *Nucl. Instrum. Meth. A*, 385:354–360, 1997.

[46] D.P. Kozlenko, B.N. Savenko, V.P. Glazkov, and V.A. Somenkov. Neutron scattering investigations of structure and dynamics of materials under high pressure at IBR-2 pulsed reactor. *Neutron News*, 16:13–15, 2005.

[47] I.N. Goncharenko. Neutron diffraction study of the equation of state of molecular deuterium at high pressure. *High Pressure Res.*, 24:193–204, 2004.

[48] K. Furuno, A. Onodera, and S. Kume. Sapphire-anvil cell for high pressure research. *Jap. J. Appl. Phys.*, 25:L646–L647, 1986.

[49] V.P. Glazkov and I.N. Goncharenko. Experiments on neutron diffraction in sapphire anvils at pressure up to 7.5 GPa. *Fizika i tekhnika vysokikh davlenii*, 1:56–59, 1991.

[50] A.M. Patselov, I.M. Demchuck, and A.A. Starostin. Pressure generation up to 15 GPa between sapphire anvils. *Pribory i tekhnika eksperimenta*, 6:157–159, 1990.

[51] W.F. Kuhs et al. Single crystal diffraction with x-rays and neutrons: High quality and high pressure? *High Pressure Res.*, 14:341–352, 1996.

[52] W.F. Kuhs, H. Ahsbahs, D. Londono, and J.L. Finney. In-situ crystal growth and neutron four-circle diffractometry under high pressure. *Physica B*, 156 & 157:684–687, 1989.

[53] D.P. Kozlenko et al. High-pressure effect on the crystal and magnetic structures of the frustrated antiferromagnet $YMnO_3$. *JETP Letters*, 82:193–197, 2005.

[54] K. Takano and M. Wakatsuki. An optical high pressure cell with spherical sapphire anvils. *Rev. Sci. Instum.*, 62:1576–1580, 1991.

[55] W.B. Daniels, M. Lipp, D. Strachan, D. Winters, and Z.-H. Yu. Simple sapphire ball cell for high pressure studies in the ultraviolet. In *Proceedings of the XIII AIRAPT - International Conference on High Pressure Science and Technology*, pages 809–811. A.K. Singh ed., Oxford & IBH Publishing Co., New Delhi, 1991.

[56] I.A. Trojan, M.I. Eremets, M.Yu. Korolik, V.V. Struzhkin, and A.N. Utjuzh. Fundamental gap of diamond under hydrostatic pressure. *Jap. J. Appl. Phys.*, 32, Suppl. 32-1:282–284, 1992.

[57] H. Sumiya and T. Irifune. Indentation hardness of nano-polycrystalline diamond prepared from graphite by direct conversion. *Diamond and Related Materials*, 13:1771–1776, 2004.

[58] M.D. Drory, R.H. Dauskardt, A. Kant, and R.O. Ritchie. Fracture of synthetic diamond. *J. Appl. Phys.*, 78:3083–3088, 1995.

[59] D. Tromans and J.A. Meech. Fracture toughness and surface energies of covalent minerals: theoretical estimates. *Minerals Engineering*, 17:1–15, 2004.

[60] C.A. Klein. Anisotropy of young's modulus and poisson's ratio in diamond. *Mat. Res. Bull.*, 27:1407–1414, 1992.

[61] J. Zhang, L. Wang, D.J. Weidner, T. Uchida, and J-A. Xu. The strength of moissanite. *Am. Mineral.*, 87:1213–1219, 2002.

[62] J. Qian, L.L. Daemen, and Y.S. Zhao. Hardness and fracture toughness of moissanite. *Diamond and Related Materials*, 14:1669–1672, 2005.

[63] E.R. Dobrovinskaia, A.L. Lytvynov, and V. Pishchik. *Sapphire - Material, Manufacturing, Applications*. Springer, 2010.

[64] M. Iwasa, T. Ueno, and R.C. Bradt. Fracture toughness of quartz and sapphire single crystals at room temperatures (in japanese). *J. Soc. Mat. Sci. Japan (Zairyo)*, 30:1001–1004, 1981.

[65] J.-A. Xu and H.-K. Mao. Moissanite: A window for high-pressure experiments. *Science*, 290:783–785, 2000.

[66] J.-A. Xu, H.-K. Mao, and R.J. Hemley. Large volume high-pressure cell with supported moissanite anvils. *Rev. Sci. Instrum.*, 75:1034–1038, 2004.

[67] Y. Ding et al. Variable pressure-temperature neutron diffracton of wüstite ($Fe_{1-x}O$): Absence of long-range magnetic order to 20 GPa. *Appl. Phys. Lett.*, 86:052505–3, 2005.

[68] T. Osakabe, H. Yamauchii, and T. Okuchi. Development of high-pressure technique for single-crystal magnetic neutron diffraction under 10 GPa. *Rev. High Pres. Sci. Techn. (in japanese)*, 20:72–75, 2010.

[69] S.P. Besedin et al. Equation of state of moleuclar deuterium at pressures up to 31 GPa. *High Pressure Res.*, 4:447–449, 1990.

[70] I.N. Goncharenko and P. Louboyre. Neutron and x-ray diffraction study of broken symmetry phase transition in solid deuterium. *Nature*, 435:1206–1209, 2005.

[71] S.P. Besedin et al. Diamond anvil cells for neutron diffraction. *High Pressure Res.*, 14:193–197, 1995.

[72] I.N. Goncharenko. New techniques for high-pressure neutron and x-ray studies. *High Pressure Res.*, 27:183–188, 2007.

[73] A. Gabke, J. Kraineva, R. Köhling, and R. Winter. Using pressure in combination with x-ray and neutron scattering techniques for studying the structure, stability, and phase behaviour of soft condensed matter and biomolecular systems. *J. Phys.: Condens. Matter*, 17:S3077–S3092, 2005.

[74] N. Vennemann, M.D. Lechner, and R.C. Oberthür. Thermodynamics and conformation of polyoxyethylene: 1. Small-angle neutron scattering and densitometric measurements at room temperature. *Polymer*, 28:1738–1748, 1987.

[75] H. Takeno et al. High pressure cell for small-angle neutron and light scattering studies of phase transitions in complex liquids. *Polym. J.*, 29:931–939, 1997.

[76] G. Pépi and P. Baroni. A new high pressure cell with sapphire windows for small-angle scattering. *J. Appl. Cryst.*, 36:814–815, 2003.

[77] M. Bonetti and P. Calmettes. Sapphire-anvil cell for small-angle neutron scattering measurements in large-volume liquid samples up to 530 MPa. *Rev. Sci. Instum.*, 76:1576–1580, 2005.

[78] J. Kohlbrecher, A. Bollhalder, and R. Vavrin. A high pressure cell for small angle neutron scattering up to 500 MPa in combination with light scattering to investigate liquid samples. *Rev. Sci. Instrum.*, 78:125101–6, 2007.

[79] H. Kohlmann, N. Kurtzemann, R. Weihrich, and Th. Hansen. In situ neutron powder diffraction on intermediate hydrides of $MgPd_3$ in a novel sapphire gas pressure cell. *Z. Anorg. Allg. Chem.*, 635:2399–2405, 2009.

[80] A.J. Rondinone, C.Y. Jones, S.L. Marshall, B.C. Chakoumakos, C.J. Rawn, and E. Lara-Curzio. A sapphire cell for high-pressure, low-temperature neutron-scattering experiments on gas hydrates. *Can. J. Phys.*, 81:381–385, 2003.

[81] N. Aso et al. A uniaxial pressure cell for neutron diffraction investigation and its use in studying the single-crystalline $Sr_3Ru_2O_7$ compound. *J. Phys.: Condens. Matter*, 17:S3025–S3028, 2005.

[82] J. Kamarád, M. Mihalik, V. Sechovský, and Z. Arnold. Miniature uniaxial pressure cells for magnetic measurements. *High Pressure Res.*, 28:633–636, 2008.

[83] B. Hälg, W. Berlinger, and K.A. Müller. Uniaxial pressure device for neutron scattering experiments. *Nucl. Instrum. Meth. A*, 253:61–64, 1986.

[84] M. Krystian, M. Prem, W. Pichl, and G. Krexner. Small-angle neutron scattering of alkali metals at moderate uniaxial pressure and low temperature. *J. Phys.: Condens. Matter*, 17:S3173–S3178, 2005.

[85] J.M. Besson, R.J. Nelmes, G. Hamel, J.S. Loveday, G. Weill, and S. Hull. Neutron powder diffraction above 10 GPa. *Physica B*, 180 & 181:907–910, 1992.

[86] S. Klotz, G. Hamel, and J. Frelat. A new type of compact large-capacity press for neutron and x-ray scattering. *High Pressure Res.*, 24:219–223, 2004.

[87] S.M. Stishov and Y.A. Sadkov. A compact hydraulic press to use with high-pressure device for neutron scattering studies. *Instruments and Experimental Techniques*, 45:573–575, 2002.

[88] H.T. Hall. Some high-pressure, high-temperature apparatus designs considerations: equipment for use at 100 000 atmospheres. *Rev. Sci. Instrum.*, 29:267–275, 1958.

[89] L.G. Khvostantsev, L.F. Vereshchagin, and A.P. Novikov. Device of toroid type for high pressure generation. *High Temperatures - High Pressures*, 9:637–639, 1977.

[90] L.G. Khvostantsev, V.N. Slesarev, and V.V. Brazhkin. Toroid type high-pressure device: history and prospects. *High Pressure Res.*, 24:371–383, 2004.

[91] C.L. Bull et al. Toroidal anvils for single-crystal neutron studies. *High Pres. Res.*, 25:229–233, 2005.

[92] W.G. Marshall and D.J. Francis. Attainment of near-hydrostatic compression conditions using the Paris-Edinburgh cell. *J. Appl. Cryst.*, 35:122–125, 2002.

[93] S. Klotz, Th. Strässle, G. Rousse, G. Hamel, and V. Pomjakushin. Angle-dispersive neutron diffraction under high pressure to 10 GPa. *Appl. Phys. Lett.*, 86:031917–3, 2005.

[94] S. Klotz et al. Magnetism and the Verwey transition in Fe_3O_4 under pressure. *Phys. Rev. B.*, 77:012411–4, 2008.

[95] M. Yousuf and K.G. Rajan. Principle of massive support in the opposed anvil high pressure apparatus. *Pramana*, 18:1–15, 1982.

[96] S. Klotz, B. Padmanabhan, J. Philippe, and Th. Strässle. The use of a 'Bridgman-seal' for low-temperature hydraulics. *High Pressure Res.*, 28:621–625, 2008.

[97] S. Klotz et al. High pressure neutron diffraction using the Paris-Edinburgh cell: Experimental possibilities and future prospects. *High Pressure Res.*, 14:249–255, 1996.

[98] E. Bourgeat-Lami et al. Overview of the projects recently developed by the advanced neutron environment team at the ILL. *Physica B*, 385&386:1303–1305, 2006.

[99] S. Klotz, Th. Strässle, A.L. Cornelius, J. Philippe, and V. Pomjakushin. Elastic properties of α-iron at high temperatures by high-pressure neutron scattering. *J. Phys. D: Appl. Phys. Phys.*, 44:055406–4, 2011.

[100] Y. Le Godec et al. Neutron diffraction at simultaneous high temperatures and pressures, with measurement of temperature by neutron radiography. *Mineral. Magazine*, 65:749–760, 2001.

[101] S. Klotz, Y. Le Godec, Th. Strässle, and U. Stuhr. The α-γ-ϵ triple point of iron investigated by high pressure - high temperature neutron scattering. *Appl. Phys. Lett.*, 93:091904–3, 2008.

[102] Y. Zhao, R.B. Von Dreele, and J.G. Morgan. A high P-T cell assembly for neutron diffraction up to 10 GPa and 1500 K. *High Pressure Res.*, 16:3424–3427, 1999.

[103] Y. Le Godec et al. A new high P-T cell for neutron diffraction up to 7 GPa and 2000 K with measurement of temperature by neutron radiography. *High Press. Res.*, 21:263–280, 2001.

[104] P.H. Fowler and A.D. Taylor. *Temperature imaging using epithermal neutrons*. Technical Report RAL-87-056, Rutherford Appleton Laboratory, Chilton, Didcot, U.K., 1987.

[105] H.J. Stone et al. Remote determination of sample temperature by neutron resonance specroscopy. *Nucl. Instrum. Meth. A*, 547:601–615, 2005.

[106] H.J. Stone et al. Temperature measurement in a Paris-Edinburgh cell by neutron resonance spectroscopy. *J. Appl. Phys.*, 98:064905–10, 2005.

[107] E. Lelièvre-Berna et al. 10 GPa precision control over a broad temperature range. *ILL Annual Report 2010*, pages 82–83, 2010.

[108] J.L. Loveday et al. Neutron diffraction studies of hydrogen-bonded ices at high pressure. *High Pressure Res.*, 17:149–155, 2000.

[109] S. Klotz et al. Techniques for neutron diffraction on solidfied gases to 10 GPa and above: Application to ND_3 phase IV. *Apl. Phys. Lett.*, 67:1188–1190, 1995.

[110] S. Ninet et al. Hydrogen bonding in ND_3 probed by neutron diffraction to 24 GPa. *Phys. Rev. B.*, 79:100101–4(R), 2009.

[111] H.E. Maynard, J.S. Loveday, S. Klotz, C.L. Bull, and T.C. Hansen. High-pressure crystallography of methane: a low-temperature neutron diffraction study in the 1-5 GPa range. *High Pressure Res.*, 29:125–128, 2009.

[112] S. Klotz, Th. Strässle, A.L. Cornelius, J. Philippe, and T.C. Hansen. Magnetic ordering in solid oxygen up to room temperature. *Phys. Rev. Lett.*, 104:115501–4, 2010.

[113] A. Bocian et al. Gas loading apparatus for the Paris-Edinburgh press. *Rev. Sci. Instrum.*, 81:093904–6, 2010.

[114] C.L. Bull et al. Note: Achieving quasi-hydrostatic conditions in large-volume toroidal anvils for neutron scattering to pressures of up to 18 GPa. *Rev. Sci. Instrum.*, 82:076101–3, 2011.

[115] Y. Wang et al. High-pressure x-ray tomography microscope: Synchrotron computed microtomography at high pressure and temperature. *Rev. Sci. Instrum.*, 76:073709–7, 2005.

[116] G.D. Bromiley, S.A.T. Redfern, Y. Le Godec, G. Hamel, and S. Klotz. A portable high-pressure stress cell based on the V7 Paris-Edinburgh apparatus. *High Pressure Res.*, 29:306–316, 2009.

[117] J. Fang et al. A rotator for single-crystal neutron diffraction at high pressure. *Rev. Sci. Instrum.*, 81:113901–6, 2010.

[118] P. Vinet, J.R. Smith, J. Ferrante, and H.R. Rose. Temperature effects on the universal equation of state of solids. *Phys. Rev. B.*, 35:1945–1953, 1987.

[119] H. Ledbetter. Relationship between bulk-moulus temperature dependence and thermal expansivity. *phys. stat. sol. (b)*, 181:81–85, 1994.

[120] Y. Wang, D.J. Weidner, and Y. Meng. Advances in equation-of-state measurements in sam-85. In *Properties of Earth and Planetary Materials at High Pressure and Temperature, Geophys. Monograph 101, M. Manghnani and T. Yagi ets.*, pages 365–372. Amercian Geophysical Union, 1998.

[121] D.L. Decker. High-pressure equation of state for NaCl, KCl, and CsCl. *J. Appl. Phys.*, 42:3239–3244, 1971.

[122] J.M. Brown. The NaCl pressure standard. *J. Appl. Phys.*, 86:5801–5808, 1999.

[123] E.F. Skelton et al. Energy-dispersive x-ray diffraction with synchrotron radiation at cryogenic temperatures. *Rev. Sci. Instrum.*, 55:849–855, 1984.

[124] O. Schulte and W.B. Holzapfel. Equation-of-state behavior for different phases of lead under strong compression. *Phys. Rev. B*, 52:12636–12639, 1995.

[125] A. Kuznetsov, V. Dmitriev, L. Dubrovinsky, V. Prakapenka, and H.-P. Weber. FCC-HCP phase boundary in lead. *Solid State Communications*, 122:125–127, 2002.

[126] H.K. Mao et al. High-pressure phase transition and equation of state of lead to 238 GPa. *Solid State Commun.*, 74:1027–1029, 1990.

[127] R.A. Miller and D.E. Schuele. The pressure derivatives of the elastic constants of lead. *J. Phys. Chem. Solids*, 30:589–600, 1969.

[128] A.D. Fortes et al. Corrigendum. *High Pressure Res.*, 32, 337 2012.

[129] D. Martínez-García et al. Equations of state of MgO at high pressure and temperature. *High Pressure Res.*, 18:339–344, 2000.

[130] P.I. Dorogokupets and A. Dewaele. Equations of state of MgO, Au, Pt, NaCl-B1, and NaCl-B2: Internally consistent high-temperature pressure scales. *High Pressure Res.*, 27:431–446, 2007.

[131] J.M. Besson et al. Variation of interatomic distances in ice VIII to 10 GPa. *Phys. Rev. B*, 1994:12540–12551, 1994.

[132] O.L. Anderson and J.E. Nafe. The bulk modulus - volume relationship for oxide compounds and related geophysical problems. *J. Geophys. Res.*, 70:3951–3963, 1965.

[133] A. Jayaraman, B. Batlogg, R.G. Maines, and H. Bach. Effective ionic charge and bulk modulus scaling in rocksalt-structured rare-earth compounds. *Phys. Rev. B*, 26:3347–3351, 1982.

[134] G. Vaitheeswaran et al. Elastic constants and high-pressure structural transitions in lanthanum monochalcogenides from experiment and theory. *Phys. Rev. B*, 75:184108–7, 2007.

[135] L. Glasser. Volume-Based Thermoelasticity: Compressibility of Inorganic Solids. *Inorg. Chem.*, 49:3424–3427, 2010.

[136] D.J. Steinberg. Some observations regarding the pressure dependence of the bulk modulus. *J. Phys. Chem. Solids*, 43:1173–1175, 1982.

[137] F. Datchi et al. Solid ammonia at high pressure: A single-crystal x-ray diffraction study to 123 GPa. *Phys. Rev. B.*, 73:094106–9, 2006.

[138] A. Dewaele, F. Datchi, P. Loubeyre, and M. Mezouar. High pressure-high temperature equations of state of neon and diamond. *Phys. Rev. B.*, 77:094106–9, 2008.

[139] W.F. Lewis, D. Benson, A.K. Crawford, and W.B. Daniels. Isochoric measurement of the equation of state of solid argon at high pressure. *J. Phys. Chem. Solids*, 35:383–391, 1994.

[140] K. Takemura. New structural aspects of elements appearing under hydrostatic pressure. In *Proceedings of the Joint 20th AIRAPT and 43rd EHPRG Conference, Karlsruhe, Germany, 2005*. ISBN 3-923704-49-6, 2005.

[141] K. Takemura. Hydrostatic experiments up to ultrahigh pressures. *J. Phys. Soc. Jpn.*, 76, Suppl. A:202–205, 2007.

[142] V.A. Sidorov and R.A. Sadykov. Hydrostatic limits of Fluorinert liquids used for neutron and transport studies at high pressure. *J. Phys.: Condens. Matter*, 17:S3005–S3008, 2005.

[143] S. Klotz, J.C. Chervin, P. Munsch, and G. Le Marchand. Hydrostatic limits of 11 pressure transmitting media. *J. Phys. D: Appl. Phys*, 42:075413–7, 2009.

[144] T. Osakabe and K. Kakurai. Feasibility test on pressure-transmitting media for single-crystal magnetic neutron diffraction under high pressure. *Jpn. J. Appl. Phys.*, 47:6544–6547, 2008.

[145] K. Murata, H. Yoshino, H.O. Yadev, Y. Honda, and N. Shirakawa. Pt resistor thermometry and pressure calibration in a clamped pressure cell with the medium, Daphne 7373. *Rev. Sci. Instrum.*, 68:2490–2493, 1997.

[146] K. Murata, S. Kagoshima, H. Yasuzuka, H. Yoshino, and R. Kondo. High-pressure research in organic conductors. *J. Phys. Soc. Jpn.*, 75:051015–14, 2006.

[147] C.A. Herbst, R.L. Cook, and H.E. King. High-pressure viscosity of glycerol measured by centifugal-force viscosimetry. *Nature*, 361:518–520, 1993.

[148] V.A. Sidorov and O.B. Tsiok. Phase diagram and viscosity of the system glycerine-water under high pressure. *Fizika i Tekhnika Vysokikh Davlenii*, 3:74–79, 1991.

[149] S. Klotz, K. Takemura, Th. Strässle, and Th. Hansen. Freezing of glycerol-water mixtures under pressure. *J. Physics: Condensed Matter*, 24:325103–6, 2012.

[150] P.W. Bridgman. The effect of pressure on the viscosity of forty-three pure liquids. *Proc. Am. Acad. Arts Sci.*, 61:57–99, 1926.

[151] K. Murata et al. Pressure transmitting medium Daphne 7474 solidfying at 3.7 GPa. *Rev. Sci. Instrum.*, 79:085101–6, 2008.

[152] R.M. Hazen and L.W. Finger. Compressibility of zeolite 4A dependent on the size of the hydrostatic pressure medium. *J. Appl. Phys.*, 56:1838–1840, 1984.

[153] K.W. Chapman, G.J. Halder, and P.J. Chupas. Guest-dependent high pressure phenomena in a nanoporous metal-organic framework material. *JACS*, 130:10524–10526, 2008.

[154] T. Sato, N. Funamori, and T. Yagi. Helium penetrates into silica glass and reduces its compressibility. *Nature Communications*, 2:345, 2011.

[155] S. Casado, H.E. Lorenzana, M. Cáceres, M. Taravillo, and V.G. Baonza. Direct measurement of the liquid 4:1 methanol-ethanol equation of state up to 5 GPa. *High Pressure Res.*, 28:637–640, 2008.

[156] R.L. Mills, B. Olinger, and D.T. Cromer. Structures and phase diagrams of N_2 and CO to 13 GPa by x-ray diffraction. *J. Chem. Phys.*, 84:2837–2845, 1986.

[157] A. Polian and M. Grimsditch. Elastic properties and density if helium up to 20 GPa. *Europhys. Lett.*, 2:849–855, 1986.

[158] M. Grimsditch, P. Loubeyre, and A. Polian. Brillouin scattering and three-body forces in argon at high pressure. *Phys. Rev. B*, 33:7192–7200, 1983.

[159] R. LeToullec, J.P. Pinceaux, and P. Loubeyre. The membrane diamond anvil cell. *High Pressure Res.*, 5:871–873, 1990.

[160] S. Klotz, J.-M. Besson, and G. Hamel. Pressure distribution in solid samples compressed by toroidal anvils. *High Pressure Res.*, 26:277–282, 2006.

[161] Ch. Maede and R. Jeanloz. Yield strength of the B1 and B2 phases of NaCl. *J. Geophys. Res.*, 93:3270–3274, 1988.

[162] S. Klotz. Phonon dispersion curves by inelastic neutron scattering to 12 GPa. *Z. Kristallogr.*, 216:420–429, 2001.

[163] G.-M. Rotaru et al. Study of diffuse scattering under hydrostatic pressure in $PbMg_{1/3}Nb_{2/3}O_3$. *J. Phys: Conf. Series*, 251:012011, 2009.

[164] P.W. Bridgman. *Solids under Pressure*. Plenum Press, New York, 1963.

[165] J.B. Parise, D.R. Locke, C.A. Tulk, I. Swainson, and L. Cranswick. The effect of pressure on the Morin transition in hematite (α-Fe_2O_3). *Physica B*, 385 & 386:391–393, 2006.

[166] D.P. Kozlenko et al. High-pressure-induced spin-liquid phase in multiferroic $YMnO_3$. *Phys. Rev. B*, 78:054401–5, 2008.

[167] I.N. Goncharenko, J.-M. Mignot, and I. Mirebeau. Magnetic diffraction studies under very high pressure. *Neutrons News*, 7(3):29–31, 1996.

[168] P. Link, I.N. Goncharenko, J.M. Mignot, T. Matsumara, and T. Suzuki. Ferromagnetic mixed-valence and Kondo-lattice state in TmTe at high pressure. *Phys. Rev. Lett.*, 80:173–176, 1998.

[169] I.N. Goncharenko, I. Mirebeau, and A. Ochiai. Magnetic neutron diffraction under pressure up to 43 GPa. Study of the EuX and GdX compounds. *Hyperfine Interactions*, 128:225–244, 2000.

[170] H. Amitsuka et al. Effect of pressure on the tiny antiferromagnetic moment of the heavy-electron compound URu_2Si_2. *Phys. Rev. Lett.*, 83:5114–5117, 1999.

[171] N. Metoki et al. Neutron scattering studies on heavy-fermion superconductors. *Physica B*, 280:362–367, 2000.

[172] H. Amitsuka et al. Pressure-temperature phase diagram of the heavy-electron superconductor URu_2Si_2. *J. Mag. Magn. Materials*, 310:214–220, 2007.

[173] O. Prokhnenko, J. Kamarád, K. Prokeš, A. Arnold, and A.V. Andreev. Helimagnetism of Fe: High pressure study of an Y_2Fe_{17} single crystal. *Phys. Rev. Lett.*, 94:107201–4, 2005.

[174] O. Prokhnenko et al. Effect of pressure and Mn substitution on magnetic ordering of $Ce_2Fe_{16}Mn$ (x=0,1). *Appl. Phys. A*, 74:S610–S612, 2002.

[175] B. Hälg, A. Furrer, and O. Vogt. Critical end point in CeSb under uniaxial pressure parallel to [001]. *Phys. Rev. Lett*, 57:2745–2748, 1986.

[176] D. Allan, W.G. Marshall, and C.R. Pulham. The high-pressure crystal structure of potassium hydrogen carbonate ($KHCO_3$). *American Mineral.*, 92:1018–1025, 2007.

[177] J.S. Loveday et al. Structure of deuterated ammonia IV. *Phys. Rev. Lett.*, 76:74–77, 1996.

[178] M.I. McMahon. Geometric effects of deuteration on hydrogen-ordering phase transitions. *Nature*, 348:317–319, 1990.

[179] C.L. Bull et al. High-pressure single-crystal neutron diffraction to 10 GPa by angle-dispersive techniques. *J. Appl. Cryst.*, 44:831–838, 2011.

[180] V.P. Glazkov et al. Experiments on neutron diffraction in sapphire anvils at pressure up to 7.5 GPa. *JETP Letters*, 47:763–767, 1988.

[181] J.-M. Mignot. Diffusion neutronique sous conditions extrêmes. *Collection SFN*, 9:159–179, 2008.

[182] H.E. Fischer, A.C. Barnes, and P.S. Salmon. Neutron and x-ray diffraction studies of liquids and glasses. *Rep. Prog. Phys.*, 69:233–299, 2006.

[183] R.J Nelmes et al. Annealed high-density amorphous ice under pressure. *Nature Physics*, 2:414–418, 2006.

[184] L. Bove, S. Klotz, J. Philippe, and A.M. Saitta. Pressure-induced polyamorphism in salty water. *Phys. Rev. Lett*, 106:125701–4, 2011.

[185] Th. Strässle et al. Structure of dense liquid water by neutron scattering to 6.5 GPa and 670K. *Phys. Rev. Lett.*, 96:067801–4, 2006.

[186] S. Klotz et al. Structure of high-density amorphous ice under pressure. *Phys. Rev. Lett.*, 89:285502–4, 2002.

[187] J.W.E. Drewitt et al. Structure of GeO_2 glass at pressure up to 8.6 GPa. *Phys. Rev. B*, 81:014202–3, 2010.

[188] Th. Strässle, S. Klotz, G. Hamel, M.M. Koza, and H. Schober. Experimental evidence for a crossover between two distinct mechanisms of amorphisation in ice I_h under pressure. *Phys. Rev. Lett*, 99:175501–4, 2007.

[189] R. Mittal, S.L. Chaplot, H. Schober, and T.A. Mary. Origin of negative thermal expansion in cubic ZrW_2O_8 revealed by high pressure inelastic neutron scattering. *Phys. Rev. Lett.*, 86:4692–4695, 2001.

[190] H. Schober and B. Renker. Pressure dependence of the external mode spectrum of solid C_{60}. *Phys. Rev. B*, 59:3287–3290, 1999.

[191] J. Eckert, W.B. Daniels, and J.D. Axe. Phonon dispersion and mode Grüneisen parameters in neon at high density. *Phys. Rev. B*, 14:3649–3663, 1976.

[192] J.W. Schmidt, M. Nielsen, and W.B. Daniels. Coherent inelastic neutron scattering study of solid orthodeuterium at high pressure. *Phys. Rev. B*, 30:6308–6319, 1984.

[193] O. Blaschko, G. Ernst, G. Quittner, W. Kress, and R.E. Lechner. Mode Grüneisen parameter dispersion relation of RbI determined by neutron scattering. *Phys. Rev. B*, 11:3960–3965, 1975.

[194] O. Blaschko, G. Ernst, and G. Quittner. Pressure induced phonon frequency shifts in KBr measured by inelastic neutron scattering. *J. Phys. Chem. Solids*, 36:41–44, 1975.

[195] Th. Strässle, A.M. Saitta, S. Klotz, and M. Braden. Phonon dispersion of ice under pressure. *Phys. Rev. Lett.*, 93:225901–4, 2004.

[196] H.A. Mook, D.B. McWhan, and F. Holtzberg. Lattice dynamics of mixed-valent SmS. *Phys. Rev. B*, 25:4321–4323, 1982.

[197] Y. Yamada et al. Lattice-dynamical properties of black phosphorus under pressure studied by inelastic neutron scattering. *Phys. Rev. B*, 30:2410–2413, 1984.

[198] A.S. Ivanov, I.N. Goncharenko, V.A. Somenko, and M. Braden. Changes of phonon dispersion in graphite at high pressure. *High Pressure Res.*, 24:193–204, 1995.

[199] Ch. Vettier and W.B. Yelon. Magnetic properties of $FeCl_2$ at high pressure. *Phys. Rev. B*, 11:4700–4710, 1975.

[200] S. Kawano, J.A. Fernandez-Baca, and R.M. Nicklow. Magnons in ferromagnetic terbium under high pressure. *J. Appl. Phys.*, 75:6060–6062, 1994.

[201] S. Klotz and M. Braden. Phonon dispersion of bcc iron to 10 GPa. *Phys. Rev. Lett.*, 85:3209–3212, 2000.

[202] Ch. Vettier, D.B. McWhan, B.I. Blount, and G. Shirane. Pressure dependence of magnetic excitations in PrSb. *Phys. Rev. Lett.*, 39:1028–1031, 1977.

[203] J. Mesot et al. Pressure-induced structural and electronic properties of high-T_c superconducting materials studied by neutron scattering. *J. Less. Comm. Metals*, 164 & 165:59–69, 1990.

[204] Th. Strässle, M. Diviš, J. Rusz, S. Janssen, F. Juranyi, R. Sadykov, and A. Furrer. Crystal-field excitations in $PrAl_3$ and $NdAl_3$ at ambient and elevated pressure. *J. Phys.: Condens. Matter*, 15:3257–3266, 2003.

[205] C. Pfleiderer, A.D. Huxley, and S.M. Hayden. On the use of Cu:Be clamp cells in magnetization and neutron scattering studies. *J. Phys.: Condens. Matter*, 17:S3111–S3120, 2005.

[206] L. Pintschovius, D. Reznik, C. Pfleiderer, and H. Löhneisen. Partial magnetic order in the itinerant-electron magnet MnSi. *Pramana*, 63:117–123, 2004.

[207] T.P. DiNoia, O.Γ. Kirby, J.H. van Zanten, and M.A. McHugh. SANS study of polymer-supercritical fluid solutions: transitions from liquid to supercritical fluid solvent quality. *Macromolecules*, 33:6321–6329, 2000.

[208] M. Bée. *Quasielastic Neutron Scattering - Principles and Applications in Solid State Chemistry*. Taylor and Francis, 1988.

[209] R. Hempelmann. *Quasielastic Neutron Scattering*. Oxford University Press, 2000.

[210] M.S. Appavou, G. Gibrat, and M.-C. Bellissent-Funel. Influence of pressure on structure and dynamics of bovine pancreatic trypsin inhibitor (BPTI): Small angle and quasi-elastic neutron scattering studies. *Biochimica et Biophysica Acta*, 1764:414–423, 2006.

[211] G.L. Squires. *Introduction to the Theory of Thermal Neutron Scattering*. Dover Publications Inc., New York, 1997.

[212] A. Furrer, J. Mesot, and Th. Strässle. *Neutron Scattering in Condensed Matter Physics*. World Scientific, 2009.

[213] H.G. Priesmeyer et al. Bragg-edge transmission as an additional tool for strain measurements. *Textures and Microstructures*, 33:173–185, 1999.

[214] V.F. Sears. Neutron scattering lengths and cross sections. *Neutron News*, 3:26–37, 1992.

[215] R.D. McCarty. Thermodynamic properties of helium 4 from 2 to 1500 K at pressure to 10^8 Pa. *J. Phys. Chem. Ref. Data*, 2:923–1041, 1973.

[216] S. Klotz, J. Philippe, and E. Cochard. Solidification and viscosity of iso-pentane/n-pentane mixtures at low temperatures and high pressure. *J. Phys. D: Appl. Phys.* , 39:1674–1677, 2006.

[217] S. Klotz, L. Paumier, G. Le Marchand, and P. Munsch. The effect of temperature on the hydrostatic limit of 4:1 methanol-ethanol under pressure. *High Pressure Res.*, 29:649–652, 2009.

Index